Jonas Firl

Probabilistic Maneuver Recognition in Traffic Scenarios

Schriftenreihe
Institut für Mess- und Regelungstechnik,
Karlsruher Institut für Technologie (KIT)

Band 031

Eine Übersicht aller bisher in dieser Schriftenreihe erschienenen Bände finden Sie am Ende des Buchs.

Probabilistic Maneuver Recognition in Traffic Scenarios

by
Jonas Firl

KIT Scientific Publishing

Dissertation, Karlsruher Institut für Technologie (KIT)
Fakultät für Maschinenbau, 2014
Tag der mündlichen Prüfung: 7. August 2014

Impressum

Scientific Publishing

Karlsruher Institut für Technologie (KIT)
KIT Scientific Publishing
Straße am Forum 2
D-76131 Karlsruhe

KIT Scientific Publishing is a registered trademark of Karlsruhe Institute of Technology. Reprint using the book cover is not allowed.

www.ksp.kit.edu

Print on Demand 2014

ISSN 1613-4214
ISBN 978-3-7315-0287-6
DOI: 10.5445/KSP/1000043680

Probabilistic Maneuver Recognition in Traffic Scenarios

Zur Erlangung des akademischen Grades

Doktor der Ingenieurwissenschaften

der Fakultät für Maschinenbau
Karlsruher Institut für Technologie (KIT)

genehmigte

Dissertation

von

DIPL.-MATH. TECHN. JONAS FIRL

Tag der mündlichen Prüfung: 07. Aug 2014
Hauptreferent: Prof. Dr.-Ing. C. Stiller
Korreferent: Prof. Dr. M. Geimer

Danksagung

In erster Linie möchte ich mich bei meinem Doktorvater Herrn Professor Stiller für seine jederzeit hilfsbereite und fundierte Betreuung meiner Arbeit bedanken. Sowohl die gegebenen wissenschaftlichen Freiheiten als auch die zahlreichen interessanten und hilfreichen wissenschaftlichen Diskussionen sind mir sehr positiv in Erinnerung geblieben. Ebenso möchte ich mich bei Herrn Professor Geimer für die Übernahme des Korreferates bedanken.

Zusätzlich möchte ich meinen Kollegen der Adam Opel AG meinen Dank aussprechen, welche mich durch meine Doktorandenzeit begleitet haben. In erster Linie möchte ich hier Norbert Simm danken, welcher die Betreuung meiner Arbeit übernommen hat und mir bei unterschiedlichsten Fragestellungen und Problemen beratend zur Seite stand. Ebenso möchte ich mich auch bei meinen ehemaligen Kollegen und Vorgesetzen bedanken, welche mich ausschließlich positiv auf meine Zeit bei Opel zurückblicken lassen. Besonders will ich hierbei meine Doktoranden-Kollegen erwähnen, welche entscheidend zu der rundum angenehmen Atmosphäre beigetragen haben.

Meinen Kollegen am MRT gilt mein besonderer Dank, da ich dort bei jedem meiner Besuche offen und freundlich aufgenommen wurde. Für das Korrekturlesen meiner Arbeit sowie für zahlreiche konstruktive und hilfreiche Hinweise bedanke ich mich darüber hinaus besonders bei Miriam Schönbein, Henning Lategahn, Quan Tran und Philipp Bender. Bei meinen Kollegen Hagen Stübing sowie Quan Tran bedanke ich mich für die konstruktive und angenehme Zusammenarbeit im Rahmen zahlreicher Veröffentlichungen. Zusätzlich möchte ich mich beim MRT-Sekretariat für die umfangreiche administrative Unterstützung sowie bei Werner Paal für den technischen Support bedanken.

Abschließend möchte ich meiner Freundin Julia meinen herzlichsten Dank sagen, da sie mich durch meine nicht immer leichte Zeit während der Promotion begleitet hat und immer an meiner Seite stand. Zu guter Letzt bedanke ich mich bei meiner Familie, im Besonderen bei meinen Eltern Doris und Siegfried, denen ich diese Arbeit widme.

Sindelfingen, im August 2014 *Jonas Firl*

Abstract

Next generation advanced driver assistance systems (ADASs) have to perform in more and more complex traffic situations including varying scenario conditions and multiple traffic participants. While current systems react mainly on single object information, future system should also be able to take different traffic participants and their relations into account. For an accurate execution of these systems a robust and accurate understanding and consideration of the entire situation is required. Thus, adequate modeling concepts have to be applied to take all spatio-temporal dependencies into account including multiple, interacting traffic participants. In this thesis an approach is presented to model and recognize traffic maneuvers in terms of interactions between different traffic participants on extra urban roads. Results of the recognition concept are presented and evaluated using different sensor setups and its benefit is outlined by an integration into a software framework in the field of Car-to-Car (C2C) communications. Furthermore, recognition results are used in this work to robustly predict vehicle's trajectories while driving dynamic traffic maneuvers.

To recognize traffic maneuvers, a probabilistic approach is proposed in this work based on Hidden Markov Models. As recognition capabilities are highly depending on the used observation data, different scenario characteristics are considered. Besides the relative kinematic information between the interacting traffic participant pair, static scenario information is added, such as about the road geometry or the type of road. Furthermore, a concept is introduced to efficiently model the influence of other static and dynamic objects by considering the required free space for the maneuver execution. The recognition framework is tested with different sequences of simulated and real world data to show the capability of the proposed method to model the traffic scene as accurate as possible.

For evaluation purposes, which is one of the most challenging tasks for situation assessment approaches, the recognition results are used to predict single traffic maneuvers like lane changes and abrupt braking maneuvers, as they are safety critical for most ADAS features. Prediction is performed by a comparison of appropriate maneuver probabilities applied to databases of more than 100 sequences for each use case. Afterwards, the points in time of the maneuver prediction are analyzed as well as their robustness in terms of true- and false-positives. Furthermore, the proposed approach is integrated into one exemplary vehicle application. Therefore, an already existing framework for verifying the trustworthiness of incoming C2C communication messages is adapted by the maneuver prediction of this thesis. Evaluations show the advantage of this approach by an increased overall security level.

Additionally, the maneuver recognition results are used to robustly predict trajectories of single traffic participants. Thus, the probabilistic information on the execution of different maneuvers is used to predict their future courses on a largest possible time horizon. The proposed concept is mainly motivated by the principles of Case-based reasoning applying a database to find similar trajectories. Results are gathered with real world data and their quality is tested by a comparison to ground truth data as well as to a state of the art predictor. Advantages of the presented method are pointed out as in predicting trajectories of single vehicles while performing dynamic, interacting maneuvers.

Kurzfassung

Zukünftige Fahrerassistenzsysteme werden in immer komplexeren Verkehrssituationen ausgeführt, wodurch unterschiedliche Anforderungen an die Systeme gestellt werden müssen. Neben einer ausreichenden sensoriellen Erfassung der Verkehrsumgebung ist besonders dessen Verstehen bzw. Interpretation von Bedeutung. Eine möglichst genaue Modellierung des kompletten Verkehrsszenarios, inklusive aller darin vorkommender, relevanter Objekte, ermöglicht es dem Anwender unterschiedliche Assistenzsysteme hierbei besonders präzise und robust ausführen zu können. Während herkömmliche Systeme mehrheitlich Informationen von einzelnen Objekten berücksichtigen, werden zukünftige Verfahren auch deren gegenseitige Einflüsse und Abhängigkeiten verstehen müssen. Die vorliegende Arbeit beschäftigt sich mit einem Teilaspekt dieses aktiven Forschungsgebietes, indem in außerstädtischen Verkehrssituationen Fahrmanövern zwischen unterschiedlichen Verkehrsteilnehmern modelliert und erkannt werden sollen.

In einem ersten Schritt wird hierfür eine geeignete Modellierungsart gewählt, welche auf einem probabilistischen Ansatz beruht. Durch die Verwendung von Hidden Markov Modellen lassen sich sowohl örtliche als auch zeitliche Abhängigkeiten zwischen einzelnen Objekten berücksichtigen, und es stehen effiziente Algorithmen für das Parametertraining und der Inferenz zur Verfügung. Um eine möglichst umfassende Modellierung des Szenarios zu ermöglichen, wird ein Verfahren vorgestellt, welches neben relevantem A-priori Wissen auch Informationen über mehrere sowohl statische als auch dynamische Objekte in das Modell integrieren kann. Das Verfahren wird mit unterschiedlichen Sequenzen bestehend sowohl aus simulierten als auch aus realen Daten getestet.

In einem nächsten Schritt wird das Verfahren evaluiert und in ein reales System integriert. Ersteres wird durch die Verwendung der Ergebnisse der Fahrmanövererkennung zur Prädiktion von sicherheitskritischen Situationen erreicht. Anhand der Prädiktion von Fahrspurwechseln und Notbremsungen für Datensätze von jeweils mehr als 100 unterschiedlicher, realer Sequenzen werden die Ergebnisse analysiert. Hierfür werden sowohl die reinen Prädiktionszeitpunkte als auch die Robustheit in Form von Falsch-Positiv-Raten erörtert. Eine Gesamtintegration erfolgt in ein bestehendes System der Sicherheit von Daten der Fahrzeug-zu-Fahrzeug Kommunikation, welches einkommende Nachrichten auf Plausibilität prüft. Die dadurch erreichte Verbesserung der Robustheit des Systems zeigt, dass das vorgestellte Verfahren für reale Fahrzeugfunktionen von großem Nutzen sein kann.

Abschließend werden die Ergebnisse benutzt, um Fahrtrajektorien von einzelnen Verkehrsteilnehmern zu prädizieren. Durch das vorhandene Wissen über die Ausführung einzelner Fahrmanöver, wird ein robustes Prädiktionsverfahren vorgestellt, welches akkurate Ergebnisse über große Zeithorizonte liefert. Basierend auf dem Prinzip des Fallbasierten Schließens wird eine Datenbasis verwendet, um aus ähnlichen Trajektorien Rückschlüsse über wahrscheinliche, zukünftige Verläufe der zu prädizierenden Trajektorie zu ziehen. Das vorgestellte Verfahren wird sowohl mit simulierten als auch mit realen Daten getestet, sowie die Ergebnisse mit Ground Truth Daten und mit den Ergebnissen eines Kalman- Prädiktors verglichen, wodurch seine Vorteile anschaulich dargestellt werden.

Contents

Abstract i

Kurzfassung iii

1 Introduction 1
1.1 Motivation 1
1.2 Contribution 5
1.3 Outline 7

2 Concepts for automated Decision Making 11
2.1 Logical Approaches 12
2.2 Probabilistic Approaches 16
2.2.1 Bayesian Networks 17
2.2.2 Dynamic Bayesian Networks 21
2.2.3 Hidden Markov Models 22
2.3 Other Approaches 28
2.4 Discussion 31

3 Probabilistic Model for Maneuver Recognition 35
3.1 Introduction 35
3.1.1 Problem Description 35
3.1.2 State of Research 37
3.2 HMM-based Maneuver Recognition 41
3.2.1 Approach 42
3.2.2 Training 43
3.2.3 Observation Model 47
3.2.4 Free Space Consideration 51
3.2.5 A Priori Knowledge 56
3.3 Maneuver Recognition Results 57

3.4 Summary 63

4 Evaluation and Application of Maneuver Recognition 65

4.1 Single Maneuver Prediction 65
4.1.1 Lane Change Prediction 67
4.1.2 Abrupt Braking Prediction 68
4.1.3 Evaluation 70
4.2 Application for Car-to-X Communications 77
4.2.1 Motivation: Car-to-X Security 77
4.2.2 Car-to-X Mobility Data Verification . . . 79
4.2.3 Results of Message Verification 83
4.3 Summary 88

5 Maneuver adaptive Trajectory Prediction 91

5.1 Motivation 92
5.2 Trajectory Prediction Approach 93
5.2.1 Case-Based Reasoning 97
5.2.2 Similarity Measure for Trajectories 99
5.2.3 CBR Cycle for Trajectory Prediction . . . 102
5.3 Results and Evaluation 109
5.3.1 Results of Maneuver dependent Prediction 109
5.3.2 Results of Prediction Approach 114
5.3.3 Evaluation 119
5.4 Summary 123

6 Conclusion 125

A Appendix 133

A.1 Sensor and Software Architecture for Observing Traffic Maneuvers 133
A.2 Integration of Message Verification Approach . . 135
A.3 Kalman Filter for Mobility Data Verification . . 137

Bibliography 141

1. Introduction

In the development of intelligent driver assistance systems in the last decades, the functionality of single features increased significantly. This may be referenced to different aspects of any ADAS (advanced driver assistance system). On a very general level, single features should provide a rapidly increased driving safety, comfort or efficiency. On the one hand, this usually requires an increased quality of the used sensor system, either by increasing the number of used sensors, by increasing the capability of single sensors or by using intelligent fusion concepts. As an example, the difference of the required sensor data for an Anti-lock braking system (ABS), one of the earliest assistance systems, and a modern active pedestrian protection system, usually requires one or multiple cameras and an extensive computational effort, is enormous. On the other hand, the situations in which the system should assist the driver become more and more complex. This requires a more accurate perception, understanding and consideration of multiple aspects of complex traffic situations. This thesis contributes to increase the information about the current traffic scene for any ADAS which is extracted from sensor data. In this introducing chapter, a motivation for this work is given in the following section as well as an overview about its contributions and outline.

1.1. Motivation

To motivate and define the scope of this thesis, a deeper look at the general information flow for driver assistance systems has to be done. In Figure 1.1 its different stages are outlined

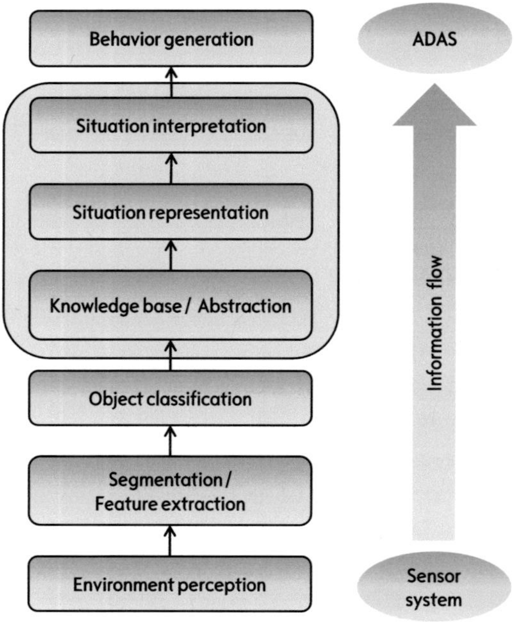

Figure 1.1.: Information flow of ADAS

simplified. From the lower levels including the sensor system and its perception of all relevant information of the traffic scenario (either information on the ego vehicle, like wheel speed or steering angle, or kinematic information about other objects) information is passed through different process units. At the highest level the behavior generation is done, which may include different interventions (automated braking, steering), driver warning strategies or only the visual or acoustic display of relevant information. Obviously, not all stages are executed similarly for all ADAS features, or are certainly not required with the same significance. Considering a Lane Departure Warning (LDW) feature, which warns the driver when leaving the own traffic lane unintentionally, a simplified abstraction and representation stage is required comparing to other features. On the other hand, most radar based features does not require significant feature extraction steps, as all relevant information are measured directly by the sensor.

As ADAS features were increased considerably in their functionality in the recent past, the situations where the feature is executed were increased in their complexity likewise. The more complex the situation is the system has to act in, the higher the requirements on the *situation interpretation* level are. These requisites have to be passed through all lower levels of the information flow. One example is the sensor level where an increased functionality on the levels above may demand new sensor types to be included or to apply more complex fusion concepts (see (Winner et al., 2009) for more detailed information about all stages of the information flow).

ADAS - State of the Art: Most of common driver assistance features are nowadays carried out as follows: A sensor system perceives the environment of the vehicle, using for example cameras or radar sensors. Performing feature extraction and classification tasks results in two different kind of information: a list of dynamic objects (other traffic participants) including their properties (e.g., position, speed, classification results) and static information about the traffic scenario (e.g., lane information, traffic signs). On the next levels this data is usually processed to obtain an adequate interpretation of the traffic scenario depending on the accessed feature. As an example consider a Forward Collision Warning (FCW) system, which warns the driver of an immediate front collision. Therefore, the received information about other vehicles including their positions is combined with information about the kinematics of the ego vehicle (mainly its predicted further trajectory). Therewith a level of risk is calculated for every single detected object, which may require an intervention of the FCW system. The most common measure therefore is a calculation of the TCC (Time To Collision). With this interpretation of the traffic scene, i.e., relevant traffic participants including their possible TTCs, an execution of the assistance system is possible accurately. Other functionalities require adapted processes certainly. However, one of the major similarities of them is the consideration of single traffic participants only. This restricts the overall system performance in situations where multiple objects do occur, which cannot be handled independently.

Traffic Maneuvers: This usually happens when traffic participants are performing particular *maneuvers*. Different interpretations or definitions are conventional therefore:

- Single objects perform traffic maneuvers like lane changes, U-turns or parking maneuvers. Thus, only a single object is involved there.
- Multiple traffic participants perform traffic maneuvers by interacting with each other. Examples therefore are overtaking or following maneuvers. In literature, this can be found exemplarily in (Gindele et al., 2010) or (Meyer-Delius et al., 2009).

The information of both of them is crucial for some assistance systems, especially as they may enable a better prediction of the further process of the traffic scene. The task of recognizing such maneuvers is highly different for both cases. Single object maneuvers are usually easier to detect, as information about only one relevant vehicle combined with static scenario properties (road geometry, lane markings) are sufficient therefore in most situations. The recognition of maneuvers in terms of interactions between different traffic participants is quite more complicated. A lot of different challenges are responsible therefore, e.g.: interactions between traffic participants include spatio-temporal dependencies which have to be considered; a lot of uncertainties and ambiguities have to be taken into account, on the input side (sensor date) and also on the output side (maneuver recognition results). In Chapter 2 a deeper look is done on this. Ambiguities about possible recognition results have to be considered, as also human experts are not always be able to classify a traffic scene to the occurrence of one single maneuver (*always* means here: at all point in time of the traffic maneuver).

The scope of this work is indicated in Figure 1.1: by recognizing traffic maneuvers in terms of interactions between multiple traffic participants a more accurate representation of the traffic scene should be accomplished. Thus a better interpretation of the scenario should be achieved. This is independent of the lower levels of the information flow, i.e., of the used sensor setup and the first processing steps of this data. Moreover, this work should be generally independent of the behavior generation level, where different ADAS features may be accessed. Furthermore, this task is highly depending on the traffic scenario, which has to be observed. As a very important criterion, inner city and outer city traffic have to be distinguished. In this work, only extra urban roads are considered.[1]

1.2. Contribution

The contributions of this thesis are manifold:

As a first step, an **appropriate modeling approach** has to be chosen to robustly recognize traffic maneuvers on extra urban

[1]Most works on situation recognition at inner city scenarios focus on specific road geometry, mostly one specific intersection. In general, the complexity in these situations is significant higher than on extra urban roads, as difficult traffic rules, multiple lanes and a variety of different object classes has to be considered.

roads. Therefore a definition of the problem has to be done, including the requirements on the model. Thus different modeling concepts have to be compared and evaluated with respect to these requirements. The selected model should consider the entire traffic environment in a best possible and realistic way. Challenges are the consideration of multiple or even arbitrary many traffic participants and multiple object classes (dynamic and static ones). Furthermore, the complexity of the complete recognition framework has to be kept feasible to be used with state of the art hardware architectures for ADAS.

As shown in Figure 1.1, the scope of this work does not depend on a specific sensor architecture. Thus, a variety of **different sensor systems** are possible to be used. As the acquired sensor data are used as input data for the modeling approach of this work, a specific quality has to be guaranteed certainly. However, the proposed concepts should be kept adaptable to different sensor platforms. To show this flexibility different kind of input sources for the proposed modeling approach should be used in this thesis. Furthermore, the focus is on state of the art hardware setups, as already been available in current vehicles, in contrast to most situation assessment approaches presented in recent years.

As one of the main challenges for all situation assessment approaches, **evaluation** cannot be done in a straight forward way. Depending on the highest level of the information flow (behavior execution), no general evaluation approach is applicable as it depends on the further processing steps and the kind of accessed ADAS feature. However, the task of evaluating the proposed approach of this work has to be treated in a way most applications may use the gained information. Usually, any kind of prediction capabilities of the overall systems are analyzed therefore.

Even if this work should be adaptable to different input sources and outputs (see Figure 1.1), its overall performance has to be presented in the context of a real **vehicle application**. As different combinations of sensors and applications are possible therefore, an appropriate overall system has to be selected. Evaluation has to be done in this context to show the benefit of the gained information about the recognized traffic maneuvers. Therefore, an integration into a complete software framework has to be done, to furthermore show the real time capabilities of the overall system.

Finally, the gained information should be used to increase the knowledge of the complete traffic scene by **predicting driving trajectories** of single traffic participants. As most driver assistance systems require robust predictions about the behavior of different traffic participants, their trajectories, especially when performing dynamic maneuvers, are of high interest. Thus, the proposed prediction approach should be able to increase robustness and prediction horizon compared to state of the art concepts. For an evaluation, the framework has to be tested with simulated date as well as with real world data.

1.3. Outline

This thesis is structured as follows:

In the following **Chapter 2** basic concepts for automated decision making systems are presented. These may be applied to the main recognition task of traffic maneuvers in this work. While considering some basic requirements already mentioned in this introducing chapter, two different groups of methods are distinguished: logical and probabilistic approaches. After giving a short overview of some logical agents in Section 2.1, the focus of this chapter are the probabilistic models of Section

2.2 and their characteristics. Therefore, Bayesian Networks (BN), Dynamic Bayesian Networks (DBN) and Hidden Markov Models (HMM) are discussed, including their definitions and possible advantages or drawbacks for the topic of this thesis. Furthermore, basic concepts of training and evaluation purposes are presented. To finish this chapter, an overview and evaluation of all concepts are outlined in Section 2.4.

The proposed probabilistic approach for recognizing traffic maneuvers is presented in **Chapter 3**. An introduction including a more detailed problem description and an overview of the relevant state of research (and its discussion) is given at the beginning in Section 3.1. The complete recognition framework is detailed in Section 3.2 describing the used configuration for parameter learning and used simulation software. The crucial information considered in the observation model is focus of the subsequent sections, as it is the indicator which maneuver characteristics may be considered by the models. The more accurate the information of the observation model is, the more reasonable the recognition capabilities are. For first evaluation purposes the recognition framework is tested and results with simulated as well as with real world data are presented.

The evaluation of the presented maneuver recognition approach is the topic of **Chapter 4**. Thus, the contribution of this chapter is twofold: Firstly, the capabilities of the recognition framework are evaluated and secondly, an example of the possible benefit for an ADAS application is presented. The first task is handled by using the recognition results for predicting typical safety critical situations in traffic scenarios. A large database of real world traffic situations is used for that purpose and different evaluation methods are conducted therewith. As an application for a specific ADAS, in Section 4.2 the proposed concepts are integrated into an already existing architecture for the security of Car-to-Car (C2C) communication. Therefore, the prediction approach is applied to adapt an already existing system for verifying the trustworthiness of C2C communication messages. Results with real world data are presented afterwards to show the increased verification capabilities.

In **Chapter 5** another possible benefit for different driver assistance systems is discussed. Using the probabilistic information of traffic maneuvers enables a more accurate and robust prediction of the trajectories of single traffic participants. Thus, an approach is presented which is motivated by the principles of Case-based reasoning (CBR). A well suited similarity measure is introduced for combining single trajectory parts with a predefined database. The prediction concept is tested with simulated and real world data again. Evaluation is done by comparing the proposed approach with ground truth data, as well as with a commonly used predictor.

Finally in **Chapter 6**, a summary of the content of this thesis is given and an outlook to ongoing work is discussed.

2. Concepts for automated Decision Making

In this chapter an overview of concepts for automated decision making is given that can be used for the recognition of traffic maneuvers. Automated decision making includes a wide field of Artificially Intelligence (AI) that cannot be discussed entirely in this thesis.

In general this task mainly addresses the problem of reasoning in a world with incomplete knowledge. In the past decades many different approaches for reasoning of artificial agents were developed to enable solving more and more complex tasks and to handle and act in more complex environments. Interested readers are referred to the works of Russell and Norvig (Russell and Norvig, 2009) and Bishop (Bishop, 2006), where an overview of different AI topics is given, and for decision theory in particular (Bather, 2000) and (Berger, 1985). Two different aspects have to be considered here: What should the concept of *knowledge representation* look like and how should *reasoning* be performed? To motivate these two questions, the scope of this work may assist: To recognize traffic maneuvers, which cannot be observed directly, the given or observable knowledge about the environment have to be taken into account, requiring an adequate knowledge representation. As a next step, reasoning about traffic maneuvers has to be performed, where usually inference algorithms are applied. In Figure 1.1, *Situation representation* and *Situation interpretation* are the two crucial stages to which these two aspects are related to.

In the following two different classes of approaches are presented with respect to their capabilities of recognizing traffic maneuvers:

- *logical approaches*, where different logical languages may be used to interpret sentences building up the *knowledge base*, and
- *probabilistic approaches*, where mainly graphical models are used to represent the joint probability distribution (JPD) over some random variables.

As traffic maneuver recognition can be interpreted as a task of automated decision making or general classification, a lot of different other approaches in the field of machine learning (Bishop, 2006) do exist. An overview of related work about the task of this thesis will be given in Section 3.1.2.

This chapter is structured as follows: In Section 2.1 automated decision making with logical agents is discussed. Section 2.2 presents probabilistic approaches afterwards, where especially Bayesian Networks (BNs), their capabilities and their extensions for maneuver recognition are introduced. Examples of another possible approach can be found in Section 2.3, where the usage of State Machines and Artificial Neural Networks (ANNs) for traffic maneuver recognition is discussed. Finally, Section 2.4 discusses the presented concepts and motivates the used approach in this thesis.

2.1. Logical Approaches

In (McCarthy, 1963a) and (McCarthy, 1963b) a first step was done for reasoning and decision making in situations using formal languages. A lot of *classical* logics were developed afterwards, see, e.g., (Harmelan et al., 2008) Chapter 1. The main component of all logic based approaches is the *knowledge base* in which the a priori knowledge about the world is stored in terms of *sentences*. To build up correct sentences the *syntax* of the language has to be considered. On the other hand the *semantic* is used to interpret these sentences. Therefore different logic languages can be used where adequate ones are presented in the following. For a discussion and evaluation please refer to Section 2.4 of this chapter.

The most basic concept for using logical agents is first order logic (FOL). Although FOL enables to formalize many different constructions from natural languages, it suffers under some basic limitations. For instance, it is not able to express any adjective and adverbial context (Vehicle A is driving *fast*, Object A is a *slow* vehicle) or any prepositions referring to time or position (Vehicle A is *next to* Vehicle B).

Modal Logic as invented by (Hintikka, 1962), extends the concept of FOL with modal operators. An introduction and its basic principles can be found in (Chellas, 1980) and (Harmelan et al., 2008) (Chapter 15). Basically, modalities that express *possibility* ($\diamond p :\Leftrightarrow p$ is possible) and *necessity* ($\Box p :\Leftrightarrow p$ is necessary) are involved as modal operators to overcome some shortcomings of FOL. Nevertheless, the term *modal logic* is used more broadly to express a variety of different modalities (e.g., temporal modalities, modalities of knowledge and belief). Modalities seem to be a very general requirement for many real world applications since they extends classical logic to model a priori knowledge, which includes uncertainties, ambiguities or temporal changes over time.

In (Bennett et al., 2002) a framework for knowledge representation and reasoning is presented using multi-dimensional (two-dimensional) modal logic. This structure is capable to represent spatial, as well as temporal relations in complex worlds (e.g., Vehicle A is behind Vehicle B). Nevertheless it has some crucial difficulties in handling complex changes over time and in keeping system complexity low, especially in worlds with increased complexity (like multiple objects with an unknown or varying number).

Temporal Logic can be considered to be a special case of modal logic, where temporal modalities are added. Examples therefore are: $\diamond p :\Leftrightarrow p$ *is true at all points in time in the future*, $\Box p :\Leftrightarrow p$ *will eventually be true at some point in time in the future.* Interested readers are referred to (Øhrstrøm and Hasle, 1995) and (Manna and Pnueli, 1992) for more detailed information on basic principles and ideas. The concept of model temporal dependencies seems to be crucial in the context of this work, as traffic scenarios contains important changes over time.

An approach for applying temporal logic for situation modeling was developed in (Schäfer, 1996) in terms of *Fuzzy Metric Temporal Horn Logic* (FMTHL). Thereby it is possible to handle both, time dependency and partial information by using temporal modalities, uncertainty and fuzzy intervals. Managing uncertain data is a very common requirement for real world applications, since data acquisition is often defective. Approaches are presented in (Munch et al., 2011), (Fernández et al., 2008) and (Gerber and Nagel, 2008) where the knowledge base, extracted from a computer vision system, is extended by using FMTHL. However, these systems still have crucial difficulties in representing complex systems, as the underlying knowledge base becomes unfeasible large and is hardly to be built up using human export knowledge. This is the case, especially when considering different probabilistic types of information.

Description Logic (DL) introduces the terminology of *concepts* for representing the knowledge base of the application domain (Baader et al., 2003). Therewith, properties of objects and individuals in the domain are described. One of the emphases of DL is the inference of knowledge from the explicit information of the knowledge base. This knowledge base is understandable due to the similarity of DL axioms to the human language, and therefore also serviceable as detailed in (Hummel et al., 2007).

In (Hummel, 2009) an approach for scene understanding based on DL is presented, more precise for intersection understanding in traffic scenarios. Therefore, some basic characteristics of computer vision are converted to a DL framework including object detection, object classification and data association. Nevertheless DL still has some general problems, typical for logic based approaches like the computational performance and the incorporation of probabilistic information which makes them unfeasible in many real world applications. See the following Section 2.4 for a discussion.

Markov Logic Networks (MLNs) as introduced in (Richardson and Domingos, 2006) represent an approach to overcome the difficult handling and consideration of probabilistic information when using logical agents. A MLN combines first-order logic with Markov random fields.[1] Therefore, weights are attached to all sentences of the first order knowledge base, which validates the assertion of the corresponding sentence. Thus, any impossible sentence is associated with 0 and all true formulas with the value 1. These values (or probabilities) have to be learned from training data, which is a non-trivial task, depending on the complexity of the modeled system.

MLNs can be applied to different application areas, such as object classification tasks (see (Bachmann and Lulcheva, 2009), (Stiller et al., 2008)) or to the recognition of objects relations

[1] Markov random fields, sometimes also called Markov Networks, are undirected graphical models.

(see (Hensel et al., 2010)). The evaluation of these approaches can be performed intuitively because of their general traceability. Although these modeling approaches seem to compensate the main drawback of many logical agents, namely the handling of uncertain information, it has some major problems in complex real world applications: The complexity of the constructed random field increases very fast with the number of objects and with the number of underlying rules which makes MLNs practically unfeasible. In addition, they generally lack in modeling temporal information and continuous random variables, as the required discretization of the input variables always implies a loss of information.

2.2. Probabilistic Approaches

After getting a brief overview of logical concepts, in the following probabilistic approaches are presented. These are graphical models, which can be roughly divided into

- Markov Networks and
- Bayesian Networks.

While a Markov Network is represented by an undirected graph, Bayesian Networks are directed acyclic graphical models. Both of them represent a joint probability distribution (JPD) over some random variables $\chi = \{X_1, \ldots, X_n\}$. Those are represented by the nodes of the corresponding graph, and their conditional dependencies by the edges (Koller et al., 2007).

One advantage of Bayesian Networks is the fact that they are easier to interpret because of their clearer probabilistic semantic. Therefore they are generally used where clear relations in terms of cause and effect are given (modeled by directed arcs in the corresponding graph). Markov Networks have their strength in representing variables with a rather correlational interaction, e.g., in tasks of image processing (Smyth, 1997).

In the following sections a deeper look at Bayesian Networks and some of their extensions is given. An overview of related work about Bayesian Networks in the field of this work can be found in Section 3.1.2.

2.2.1. Bayesian Networks

Bayesian Networks (BNs) are a useful tool to represent the JPD over a set χ of n random variables X_i. The corresponding acyclic directed graph G represents conditional independencies, so that each variable X_i is independent of its nondescendents given its parents $par(X_i)$ (Markov property). The JPD of χ is then given as:

$$P(x_1,\ldots,x_n) = \prod_{i=1}^{n} P(x_i|par(X_i)), \qquad (2.1)$$

where $P(x_i)$ is used as an abbreviation for $P(X_i = x_i)$. When defining Θ_i $(i = 1,\ldots,n)$ as the set of parameters for representing the conditional probability distribution of X_i given its parents, a BN can be defined as the tuple $B := (G,\Theta)$.

The direct representation of conditional independencies reduces the model complexity of the JPD in many cases significantly. That means, the number of required model parameters are reduced whatever the random variables are discrete or continuous valued. In the simpler, discrete case a specification of the conditional probabilities can be done for each possible value. However, in most real world scenarios continuous quantities were observed, which have infinite possible values. In these cases two possible approaches may be applied: a discretization of the continuous value range into predefined intervals, which comes along with a loss of information and accuracy. Another way is to handle continuous variables is the usage of parametrized probability density functions as approximations, e.g., the Gaussian distributions $\mathrm{N}(\mu,\sigma^2)(x)$ with 2 parameters

$\Theta = (\mu, \sigma)$. In the following the two main tasks for BNs (as well as for all graphical models) are discussed: learning the model and evaluating inference queries.

Learning in BNs: In most applications the setting of the BN is unknown and has to be learned from training data D. This may imply two different aspects: the network structure G and the parameters Θ of the JPN (2.1). The first task, learning the structure of the network, is a very challenging task which is often performed by using expert knowledge, especially if the complexity is assumed to be restricted. To solve this problem analytically, definitions of the search space of possible network structures, as well as a measure or score function are required. The score function may be defined by using a posterior distribution $P(G|\mathrm{D})$ of the possible network structures given the training data set and can be approximated by sampling over the structures using Markov chain Monte Carlo (MCMC) methods. See (Friedman and Koller, 2003) for more detailed information. For learning the parameters Θ of the JPD, it is often assumed that the underlying model structure is known (Ben-Gal, 2007). Therefore, the learning task is to find the model parameters that match best with the training data. The basic element therefore is the likelihood function, i.e., the probability of the training data $\mathrm{D} = \{\mathbf{x_1}, \ldots, \mathbf{x_m}\}$, with $\mathbf{x_i} = (x_{i1}, \ldots, x_{in})^T$, given the model, where usually the log-likelihood function is used instead:

$$\log P(\Theta|\mathrm{D}) = \sum_{j=1}^{m} \sum_{i=1}^{n} \log P_{\Theta_i}(x_{ji}|par(X_i)). \tag{2.2}$$

The maximization of this function over all possible model parameters Θ is the Maximum Likelihood solution for BN model parameter estimation. Other approaches, e.g., Bayesian parameter estimation can be found in (Koller et al., 2007).

Inference in BNs: If the graphical structure as well as the parameters of a BN are given, different inference queries may be evaluated. In general, inference in BNs is the task to estimate the posterior probability distribution $P(\mathbf{X}|\mathbf{o})$ of one or more variables $\mathbf{X}$ for a given (i.e., observed) event $\mathbf{o}$ of some evidence variables $\mathbf{O}$ and some non-evidence (i.e., hidden) variables $\mathbf{Y}$, such that:

$$\chi = \mathbf{X} \cup \mathbf{O} \cup \mathbf{Y}. \tag{2.3}$$

An exact computation of $P(\mathbf{X}|\mathbf{o})$ requires in a straight forward way the summation over all non-evidence variables, as a worst case scenario:

$$P(\mathbf{X}|\mathbf{o}) = \alpha \sum_{y \in \mathbf{Y}} P(\mathbf{X}, \mathbf{o}, y), \tag{2.4}$$

where α is a normalization constant. Even for Boolean variables, the computational complexity is $O(n2^n)$, by what this approach is intractable for large networks. For specific network structures, there exists some algorithms for effectively compute the probability in Eq. (2.4), like variable elimination for polytrees (single connected networks). However, for general networks, exact inference remains a NP-hard problem, see (Russell and Norvig, 2009) for more information.

Therefore, approximative inference algorithms are used in most applications based on sampling techniques. The simplest way to sample is direct sampling, where events $x_1, \ldots, x_n$ are generated by sampling each variable in a topological order. If this procedure is performed N times and let denote the frequency of a specific event $(x_1, \ldots, x_n)$ with $N(x_1, \ldots, x_n)$ and the probability to sample this event with $P_S(x_1, \ldots, x_n)$, the following equation holds:

$$\lim_{N \to \inf} \frac{N(x_1, \ldots, x_n)}{N} = P_S(x_1, \ldots, x_n) =$$
$$= \prod_{i=1}^{n} P(x_i | par(X_i)) = P(x_1, \ldots, x_n) \quad (2.5)$$

Hence, inference can be performed by rejecting those samples which do not match to the given evidence **o**. In real world applications this approach is often unfeasible, because the number of rejected samples increases exponentially with the number of evidence variables. Likelihood weighting may be applied to compensate this drawback, but nevertheless its performance still depends on the number of evidence variables (and furthermore on the occurrence of the ordered variables, see (Russell and Norvig, 2009)). Besides these approaches, MCMC algorithms are very useful for probabilistic sampling, see (Andrieu et al., 2003). MCMC changes only one variable in each step but keeping the evidence variables fixed. Transition probabilities are calculated with different sampling approaches, like the Gibbs sampler.

BNs, as well as some of their extensions like object-oriented Bayesian Networks (OOBN, see (Kasper et al., 2012)), provide an approach for automated decision making and reasoning in situations with incomplete knowledge in a probabilistic manner. Hence, they are capable to deal with uncertain and incomplete knowledge, which is a prerequisite for most real world applications. Nevertheless, they have some major drawbacks in handling dynamic, temporal dependencies and in efficiently solving exact inference tasks, especially in large, multiply connected networks.

2.2.2. Dynamic Bayesian Networks

Dynamic Bayesian Networks (DBNs), first applied in (Dean and Kanazawa, 1989), extends the theory of Bayesian Networks to handle temporal probability models and sequential data. The main difference to BNs is that not only a set of (unobservable, hidden) state variables $\mathbf{X}$ and (observable) evidence variables $\mathbf{O}$ are considered, but a temporal model is applied: $\mathbf{X}_t$ denotes the set of state variables at time t and $\mathbf{O}_t$ denotes the evidence variables at time t. In most cases it is assumed, that the sets of state and of evidence variables do not change over time.

For defining a DBN, three kinds of information have to be considered: The initial distribution over the system states $P(\mathbf{X}_0)$, the transition model for the system states $P(\mathbf{X}_t|\mathbf{X}_{0:t-1})$ and the observation model (or sensor model) $P(\mathbf{O}_t|\mathbf{X}_{0:t}, \mathbf{O}_{0:t-1})$, describing how the observation is affected by previous variables. These values have to be specified for constructing a DBN, what is unfeasible because of the unbounded number of required parameters. Therefore two assumptions are usually made:

- **Markov assumption:** This assumption restricts the number of parents for the system variables and the observation variables by making the current state only dependent on a finite set of previous states. When using the simplest case of this assumption, namely the first order Markov assumption, the following equations hold:

$$P(\mathbf{X}_t|\mathbf{X}_{0:t-1}) = P(\mathbf{X}_t|\mathbf{X}_{t-1}) \tag{2.6}$$

$$P(\mathbf{O}_t|\mathbf{X}_{0:t}, \mathbf{O}_{0:t-1}) = P(\mathbf{O}_t|\mathbf{X}_t) \tag{2.7}$$

- **Stationarity:** The conditional probability for state transitions does not change over time, so Eq. (2.6) and Eq. (2.7) hold for all t.

Inference in DBNs: The task of evaluating inference queries in DBNs is similar to the one in BNs, as they can be transformed to normal BNs by duplicating the time slices until the network is large enough for the current observation data. This procedure is called unrolling and enables applying the same exact inference algorithms as named in Section 2.2.1. Unfortunately, this results in the same problematic complexity, especially for large number of variables. Also approximate inference algorithms may be applied, which can be roughly divided into stochastic and deterministic approaches, see (Murphy, 2002) for a detailed analysis.

Even if DBNs are capable of handling temporal systems, some drawbacks of BNs still exist, like the difficult learning of the network structure, in particular for complex systems, which require human expert knowledge in most cases. Furthermore inference still remains a challenging task for exact queries in complex networks.

2.2.3. Hidden Markov Models

Hidden Markov Models (HMMs) are special cases of DBNs for probabilistic reasoning in temporal systems. As first introduced in (Baum and Petrie, 1966), they are extensively used for applications in speech recognition (see (Rabiner, 1989)). In the last decades their application areas were extended to different classification and recognition tasks, some of them can be found in (Zucchini and MacDonald, 2009).

A HMM can be introduced as a DBN with only one discrete state variable X_t and one observation variable O_t, forming two stochastic processes. Assuming there are n possible states $x_1, \ldots, x_n$ for X_t and Eq. (2.6) and Eq. (2.7) hold for all t, a HMM is defined by the following:

- The probabilities of the transitions from state x_i to state x_j defines the *state transition matrix* $A = \{a_{i,j}\}$ $(i,j = 1,\ldots,n)$, with:

$$a_{i,j} := P(X_t = x_j | X_{t-1} = x_i) \quad \forall i,j = 1,\ldots,n. \tag{2.8}$$

- The *observation model* $B = \{b_i\}$ $(i = 1,\ldots,n)$ specifies the probability of making an observation o while being in state x_i:

$$b_i := P(O_t = o | X_t = x_i) \quad \forall i = 1,\ldots,n. \tag{2.9}$$

- The *initial state distribution* $\pi = \{\pi_i\}$ $(i = 1,\ldots,n)$ defines the probabilities of being in state x_i at the initial point in time $t = 0$:

$$\pi_i := P(X_0 = x_i) \quad \forall i = 1,\ldots,n. \tag{2.10}$$

A HMM is thus defined by the triple $\lambda = (A, B, \pi)$. The finite dimensions of the state transition matrix $A \in \mathbb{R}^{n \times n}$ and of the initial state distribution $\pi \in \mathbb{R}^n$ are defined by the number of discrete system states n. But otherwise, the dimension of the observation model B depends on the number of different observations, and on whether discrete or continuous values are observed. For the simple, discrete case with m observation symbols $v_1,\ldots,v_m$, B can be written as an $m \times n$ matrix with

$$b_{j,i} := P(O_t = v_j | X_t = x_i). \tag{2.11}$$

For continuous observations the same approaches as noted in Section 2.2.1 for BNs may be applied. When using an approximation with probability density functions, the dimension of B is further increased, which implies an increased number of model parameters. The structure of one HMM with

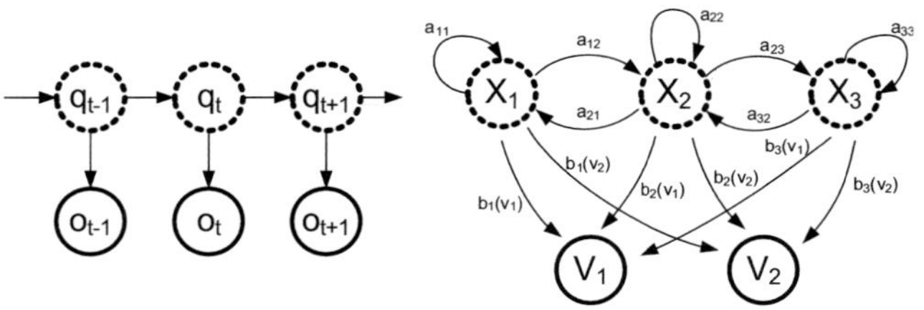

Figure 2.1.: Structure of a HMM with three system states and two observation symbols.

three system states and two discrete observation symbols can be exemplarily seen in Figure 2.1.

The assumptions from Eq. (2.6) and Eq. (2.7) can be rewritten for HMMs to:

$$P(X_t = x_i | X_{0:t-1}) = P(X_t = x_i | X_{t-1}) \quad \forall i = 1, \ldots, n \tag{2.12}$$

$$P(O_t = v_j | X_{0:t}, O_{0:t-1}) = P(O_t = v_j | X_t) \quad \forall j = 1, \ldots, m \tag{2.13}$$

Similar to BNs, the training of the model structure and parameters as well as applying inference queries are the two major tasks when working with HMMs.

Learning in HMMs: For learning the structure of HMMs different approaches may be applied, as possible structures of the network are often restricted by expert knowledge. Some popular kinds of HMMs can be found in (Bengio, 1999). Focusing on the task of learning the model parameters λ for a given training set $\mathbf{o}_{train} = (o_0, \ldots, o_T)$ usually means solving the maximization problem:

$$\hat{\lambda} = \arg\max_{\lambda} P(\mathbf{o}_{train} | \lambda). \tag{2.14}$$

Since there does not exist an analytical solution to this problem, approximate, iterative approaches are applied yielding in a local optimization of $P(\mathbf{o}_{train}|\lambda)$. The most common one is the *Baum-Welch algorithm* or the more general *EM method* (expectation-maximization). It can be roughly described by the following steps (Zucchini and MacDonald, 2009):

1. Select the initial parameters $\lambda_0 = (A_0, B_0, \pi_0)$.

2. **E-step:** Compute the expectations for the model parameters given the observation sequence $\mathbf{o}_{train}$ and the last estimate of λ_t. Thus, it is necessary to compute the following:

   - Expected number of state transitions from state x_i given the current model parameters and given the observation sequence:

   $$\sum_{t=0}^{T-1} \sum_{j=1}^{N} P(X_t = x_i, X_{t+1} = x_j | \lambda, \mathbf{o}_{train}).$$

   - Expected number of state transitions from state x_i to state x_j given the current model parameters and given the observation sequence:

   $$\sum_{t=0}^{T-1} P(X_t = x_i, X_{t+1} = x_j | \lambda, \mathbf{o}_{train}).$$

3. **M-step:** Maximize $P(\mathbf{o}_{train}|\lambda_t)$, the conditional probability with respect to λ, using the calculated expectations, yielding in new model parameters λ_{t+1} satisfying:

   $$P(\mathbf{o}_{train}|\lambda_t) \leq P(\mathbf{o}_{train}|\lambda_{t+1}) \qquad (2.15)$$

4. Repeat point 2 and 3 until a predefined convergence criterion has been satisfied.

A detailed description of the Baum-Welch algorithm can be found in (Rabiner, 1989). It converges to a local maximum based on its initial parameters, like in other non-convex problems. Therefore, λ_0 has to be chosen wisely to find robust and stable results. Furthermore, multiple runs of the complete algorithm are usually done.

Inference in HMMs: Inference tasks in HMMs may be divided into two following groups:

- Evaluation: For a given HMM λ and a given observation sequence $\mathbf{o} = (o_1, \ldots, o_T)$ the conditional probability $P(\mathbf{o}|\lambda)$ has to be computed.
- Decoding: For a given HMM λ and a given observation sequence $\mathbf{o} = (o_1, \ldots, o_T)$ the corresponding (hidden) state sequence $\mathbf{X} = (X_1, \ldots, X_T)$ has to be chosen.

Compared to inference in general BNs/DBNs, HMMs provide exact inference algorithms for both tasks, due to their simplified structure.

Evaluation in a straight forward way is unfeasible, since its complexity is $O(TN^T)$. The so called *Forward algorithm* enables an effective evaluation for HMMs, due to their structure. The *Forward variable* $\alpha_t(i)$ for a given point in time t and state x_i is the conditional probability of observing the subsequence $(o_1, \ldots, o_t)$ ending in state $X_t = x_i$ given the model parameters λ:

$$\alpha_t(i) := P(o_1, \ldots, o_t, X_t = x_i|\lambda) \tag{2.16}$$

The Forward algorithm is then recursively defined, as shown in Algorithm 1. The complexity of computing the conditional probability $P(\mathbf{o}|\lambda)$ using this procedure is $O(n^2T)$, which makes it feasible, even for large observation sequences $\mathbf{o}$, as it only increases linearly with its length T.

Algorithm 1 Forward algorithm

Require: model parameter (A, B, π)
  for all i=1,...,n **do**
    $\alpha_1(i) = \pi_i b_i(o_1)$
  end for
  for all t=1,...,T-1 **do**
    for all j=1,...,n **do**
      $\alpha_{t+1}(j) = (\sum_{i=1}^{n} \alpha_t(i) a_{i,j}) b_j(o_{t+1})$
    end for
  end for
  $P(\mathbf{o}|\lambda) = \sum_{i=1}^{n} \alpha_T(i)$

Decoding of a given observation sequence $\mathbf{o}$ requires an exact definition of an "optimal" state sequence to $\mathbf{o}$. The most common optimization criterion therefore is to find the state sequence $\hat{\mathbf{X}}$ that maximizes $P(\mathbf{X}|\mathbf{o}, \lambda)$, i.e., that maximizes $P(\mathbf{X}, \mathbf{o}|\lambda)$[2]. This problem can be solved by using the *Viterbi algorithm*, a dynamic programming method. There, the highest probability $\delta_t(i)$ for a single state sequence $X_1, \ldots, X_t$ ending in $X_t = x_i$ and the first t observations is defined as:

$$\delta_t(i) := \max_{X_1, \ldots, X_{t-1}} P(X_1, \ldots, X_{t-1}, X_t = x_i, o_1, \ldots, o_t|\lambda) \quad (2.17)$$

The Viterbi algorithm is then recursively defined, as shown in Algorithm 2. As for the evaluation task, the Viterbi algorithm provides an analytical decoding procedure with a complexity of $O(n^2 T)$.

HMM based approaches highly benefit from their well-established theory about exact inference algorithms. Some of their

[2] Another possible optimization criterion may be defined by choosing each state X_t, which is individually most likely. This may result in non-valid state sequences, see (Rabiner, 1989).

Algorithm 2 Viterbi algorithm

Require: model parameter (A, B, π)
  for all i=1,...,n **do**
    $\delta_1(i) = \pi_i b_i(o_1)$
    $\Psi_1(i) = 0$
  end for
  for all t=2,...,T **do**
    for all j=1,...,n **do**
      $\delta_t(j) = \max_{1 \leq i \leq n}(\delta_{t-1}(i) a_{i,j}) b_j(o_t)$
      $\Psi_t(j) = \arg\max_{1 \leq i \leq n}(\delta_{t-1}(i) a_{i,j})$
    end for
  end for
  $\hat{X}_T = \arg\max_{1 \leq i \leq n}(\delta_T(i))$
  for all t=T-1,...,1 **do**
    $\hat{X}_t = \Psi_{t+1}(\hat{X}_{t+1}$
  end for

advantages and disadvantages can be furthermore found in (Bilmes, 2006) and (Bilmes, 2004), respectively. One of their major drawbacks is the necessity of either accurate human expert knowledge or employing a large set of training data. Furthermore, prediction approaches using the underlying model is not possible in a straight forward way (discussed in Chapter 5, see also (Omerbegovic and Firl, 2013)). However, HMMs seem to be very well suited for automated decision making in many areas, as they provide a completed set of efficient algorithms and enable flexible network structures.

2.3. Other Approaches

In addition to the logical and probabilistic concepts for automated decision making described in the last sections, there exist other approaches, which may be applied to the task of this work. Two of them are briefly discussed in the following.

State Machines yield a simple approach to model a system behavior that changes from one situation to another. Mainly used to model robotic behavior, they are capable to engage and control specific traffic maneuvers. A finite state machine is defined by

- a set of states (represented by the nodes of the corresponding graph), where only 1 node is active at any point in time,
- a set of inputs,
- a transition functions to a new state given the actual state and the given input,
- an initial state,
- a set of terminal states.

In (Kammel et al., 2008), a state machine (more precise an *hierarchical state machine*, HSM) is used to implement an intelligent maneuver planer on behavior generation level to perform autonomous driving maneuvers. The basic concept of HSMs is to group all states such that all sub-states are specializations of the corresponding parent state. The desired advantage is that the number of required state transitions is reduced so it becomes easier to capture the complex system's behavior. All possible behaviors of the vehicle are modeled as a single state, which consists of several sub-states. As an example, the state *Zone* consists of several sub states of performing different maneuvers on parking zones.

State machines represent an easy way to invoke different traffic maneuvers based on a predefined set of situational states. However, even for situations of small overall complexity, difficult state transitions and often a large number of system states are required. This often results in unfeasible network structures, in which inference algorithms are only hard to be performed. Also the assumptions that only one system state is allowed to be active at any given point in time seems to be too much restrictive, as common traffic scenarios have usually a probabilistic nature.

Artificial Neural Networks (ANNs), motivated by neural networks as used for modeling networks of biological neurons, are another possible approach for building up models for automated decision making. A possible introduction to basic concepts may be found in (Russell and Norvig, 2009). Its smallest unit is a *neuron*, which consists of multiple inputs including corresponding weights. Each neuron emits an output signal, depending on the sum of its weighted inputs and on the underlying emission function. The structure of an ANN is defined by the number of neurons, their spatial alignment and all required connections. Basic network structures are *feed-forward-* and *feed-backward-networks*. Training of ANNs may be performed either by supervised learning algorithms (e.g., Delta rule or backpropagation) as mainly used for recognition and classification tasks, or by unsupervised learning algorithms (e.g., Hebbian learning) as used for prediction issues.

Automotive applications of ANNs, especially in the field of traffic situations and their assessment, are presented exemplary in (Selinger, 2009) and (Keßler, 2011). In the latter, a master thesis supervised by the author of this work, a special case of feed-forward neural network is used, named Cascade-forward network, where every layer of neurons is connected to the input layer as well. Therewith, a recognition framework is built up for traffic maneuvers in terms of interactions between single vehicles. While learning is done with simulated data only, evaluation of the trained networks is done with simulated data as well as with real world data.

Although ANNs becomes more and more popular in the last decades for a wide variety of application areas, they still suffer from some major drawbacks. The first one is the challenging training stage of the network. Therefore a specific structure has to be determined, which is usually done empirically. There exist some well-established rules for selecting network layouts for a given problem. However, different network types have to be implemented and evaluated with respect to the given problem in most cases. Another problem is the very restricted readability of ANNs, as they are characterized as *black box* devices. Thus, the optimization or adaption of any given network becomes nearly impossible, since single model parameters cannot be interpreted directly. So, any well suited characteristic is not reproducible or guaranteed when building up new networks.

2.4. Discussion

In this section a discussion of the approaches presented in Section 2.1 and Section 2.2 is done and their usefulness for recognizing traffic maneuvers in this work is evaluated. Therefore the requirements on this concepts have to be considered, as already briefly noticed in Chapter 1:

- Traffic situations usually involve **multiple objects**, or rather objects with an unknown or varying number. For most approaches this results in complex models making inference algorithms quite difficult and resource-consuming. Thus, the proposed approach has to consider multiple objects while keeping generally applicable concerning its computational costs.

- Objects in traffic situations have **spatial-temporal dependencies**. Especially the consideration of the temporal component, i.e., the changes of object's states and their relations over time, is a prerequisite for robust modeling approaches.

- The input (observation) data is usually **uncertain**, dependent on the applied sensor architecture. These uncertainties have to be taken into account by the selected model.
- The results when recognizing traffic maneuvers also need to be somehow 'probabilistic', i.e., the identification of different traffic maneuvers is not definite. Also human experts cannot identify each situation exactly, so uncertainties and **ambiguities** have to be considered.

Most logic based approaches are powerful tools for the representation of and the reasoning in real world scenarios. With some of their modal operators they are capable to handle systems with temporal dependencies and variables changing over time. But nevertheless, the modeling of uncertainties and ambiguous as well as the generation of probabilistic results are hardly possible, what is a prerequisite in this work. Furthermore, they seem to be somehow unfeasible for complex situations due to an extended knowledge databases, e.g., in situations with multiple objects. Markov Logic Networks attempt to extend logical concepts in a probabilistic manner to handle uncertainties, but still have drawbacks when cooperating with continuous or temporal data.

In this chapter probabilistic approaches were discussed in terms of Bayesian Networks and some of their extensions. While they are capable of probabilistic reasoning and inference, no temporal dependencies could be taken into account. Dynamic Bayesian Networks compensate this drawback, usually resulting in expanded networks, making efficient inference algorithms hardly performable. Also parameter learning remains challenging due to their complex network structures. Hidden Markov

Models overcomes this problem by restricting the general network structure resulting in efficient learning algorithms and the possibility of performing exact inference.

In Table 2.1 a summary of the presented approaches is presented, including their major capabilities to satisfy important requirements for the scope of this thesis.

	Multiple objects	Temporal dependencies	Uncertainties Ambiguities	Inference Training
Logical Approaches	-	o	-	o
Markov Logic Networks	-	-	o	o
Bayesian Networks	o	o	+	o
Dynamic Bayesian Networks	-	+	+	o
Hidden Markov Models	o	+	+	+

Table 2.1.: Approaches for automated decision making and their capabilities to satisfy different requirements for situation recognition in traffic scenarios.

3. Probabilistic Model for Maneuver Recognition

After having an overview of different concepts for automated decision making, in this chapter the approach for probabilistic modeling of traffic maneuvers is presented. This approach is mainly motivated by the discussion of the previous chapter, where assets and drawbacks of different concepts were outlined.

Before describing the maneuver recognition framework, an introduction is given in Section 3.1, where the problem is formulated in more detail and the state of research for maneuver recognition approaches is briefly presented. After that, the proposed modeling approach is introduced in Section 3.2 with all relevant stages. Recognition results with simulated as well as with real world data are presented afterwards in Section 3.3. The chapter is finally summarized in Section 3.4.

3.1. Introduction

In this section the problem of modeling and recognizing traffic maneuvers is described. First, a precise description of the problem to be solved and a summary of the current state of research are given.

3.1.1. Problem Description

The problem of recognizing traffic maneuvers in terms of interactions between different traffic participants was already introduced in Chapter 1 (Figure 1.1). Current ADASs operate based on information of single objects in the vehicle's environment.

(a) Intersection: Mühlburger Tor, Karlsruhe.

(b) Autobahn A5, Karlsruhe.

Figure 3.1.: Two different traffic scenarios, with different characteristics and challenges for maneuver recognition approaches.

This information is sufficient for many current applications. For future systems, especially when going the next steps towards autonomous driving, a more accurate recognition and consideration of the vehicles surrounding are required. The relationships and interactions between different vehicles are examples of these challenges.

Interactions between traffic participants are highly depending on the current scenario. While driving on urban streets, more additional constraints have to be taken into account than on extra urban streets, like multiple traffic lights, one-way streets, impasses and pedestrians/cyclists. This work focuses on extra urban situations, like rural roads or highways. In Figure 3.1 an inner city scenario and a highway situation are depicted exemplary to show different challenges for maneuver recognition, e.g., complex traffic rules (varying right of ways at different types of intersections), traffic lights or multiple classes of traffic participants (pedestrians, bikes, motorcycles, cars, trams).

Thus, the task of maneuver recognition in the context of this thesis can be described as follows:

Maneuver recognition: Traffic maneuvers in terms of interactions between different traffic participants have to be recognized in a probabilistic manner, depending on the current kinematic information of all relevant objects in the traffic scene (e.g.,: position information, speed, acceleration) and the traffic scenario (e.g.,: lane information, road boundaries, static objects), if available. It should not depend on the source of input data (sensor system) and on the application (ADAS) that makes use of this information (see Figure 1.1).

3.1.2. State of Research

Recently, many different approaches for maneuver recognition have been developed focusing on different situations and handling different driving maneuvers. In the following a summary of different works using probabilistic models is presented including their limitations. These approaches are applied for maneuver recognition tasks to inner city as well as to extra urban traffic scenarios, as similar probabilistic approaches may be applied to both of them.

In (Schubert and Wanielik, 2011) a framework is presented to handle the assessment of objects and situations including their uncertainties in extra urban traffic scenarios for detecting lane change maneuvers. Therefore, Bayesian Networks are used to consider the uncertainties of perceived vehicles and lane markings. The corresponding network consists of different discrete nodes which represent information about all lanes (i.e., whether the lanes are either *free*, *occupied* or potentially *dangerous*) or the feasibility of lane change maneuvers (i.e., whether the maneuver is *possible*, *impossible* or *safe*). The output of the network is the *decision node* containing the values *Keep lane*, *Lane Change Left/Right*, which can be inferred exactly due to the small number of discrete nodes.

However, some of the drawbacks of BNs mentioned in Section 2.2 can be noticed in this work: The estimation of the model parameters, i.e., the definition of the conditional probability tables, requires human expert knowledge to take all relevant rules for lane change maneuvers into account. Furthermore, no temporal dependencies of the objects are considered and the interactions of traffic participants are not modeled explicitly. Besides that, the general evaluation of this interesting approach is not conducted, but only one sequence of simulated and one sequence of real data is discussed, respectively. That is why no statistical conclusion is done, like the analysis of falsely recognized lane changes (false positives).

The goal of the approach presented in (Kasper et al., 2012) is the detection of lane change maneuvers. It uses *object-oriented Bayesian Networks* (OOBNs) (Koller and Pfeffer, 1997), which extend BNs by using instance nodes to represent network segments within one single node. Results are read out from one single decision node again, which has discrete values like *Following Object*, *Following lane* or *Cut in/out*. The overall framework is easy to read and to interpret and was evaluated with a set of real world maneuvers.

Like classical BNs, OOBNs are not capable of modeling temporal dependencies in the system, so that the authors already thought about the usage of Dynamic Bayesian Networks. In addition no relative kinematic information between the traffic participants, like the relative distance, speed and acceleration is included, what is a clear potential for improvement of the proposed approach. The consideration of information of the surrounding of the ego vehicle is also an open issue, so that the framework achieves only reasonable results in simplified scenarios, where no objects like construction sides or other static objects appear.

In (Liebner et al., 2012) and (Liebner et al., 2013) another probabilistic framework for the recognition of traffic maneuvers based on BNs is outlined, which is applied for a right turn assist at inner-city intersections. The three nodes of the network are the driver's intention I (e.g., *go straight*, *turn right but stop*), the applicable driving hypothesis H, clustered from the parameters of the proposed *Intelligent Driver Model* (IDM)[1] and the observation O. The calculation of the required probability $P(I|O)$ is done by summing up over all H_i, what is applicable due to the small number of 24 applied hypotheses. Measurement data was acquired at one intersection using different drivers and different traffic conditions (with and without a preceding vehicle).

The proposed approach shows reasonable results with real world data and evaluations of prediction accuracy are given. Nevertheless, this approach is restricted to the given intersection geometry and the generalization to arbitrary scenarios is still an unsolved task (see also Section 1.1 of this thesis). In more complex situations the precise modeling and consideration of the interacting behavior between different traffic participants will become more challenging, but this issue is not handled and integrated so far by the authors.

The authors of (Gindele et al., 2010) use Dynamic Bayesian Networks to estimate the behavior of traffic participants as well as their future trajectories. The variables of the network represent the vehicle states (e.g., position and speed), situational features (e.g., $d_{long,f}$ longitudinal distance to vehicle ahead, $d_{long,b}$), recognized situation classes (e.g., *vehicle is close* and *vehicle is far away*), behaviors (e.g., *free ride* and *sheer in*), trajectories (interpreted as the realizations of the correspond-

[1] In general, the IDM is an approach for continuous modeling driving following maneuvers. In the above mentioned approach the model is extended to account for decelerations occurring on right turn situations.

ing behaviors) and a measurement vector, which is assumed to be a subset of the vehicle states. All variables are interpreted as vectors of the size of the number of traffic participants in the scene. The dynamics of the system are modeled as a first order Markov process to keep model complexity low. The approach is tested with simulated data with two cars on a two-lane road performing an overtaking maneuver with one lane change.

The proposed framework is able to handle temporal dependencies between traffic participants in a simplified scenario at a relatively low model complexity. More complex situations (e.g., more objects or urban environments) would result in more complex model structures and a highly increased computational effort for performing inference algorithms. The used joint probability function including the conditional independencies of the variables is set by human expert knowledge as well as the conditional probability functions, which is a quite error-prone procedure only working in straightforward scenarios.

In (Meyer-Delius et al., 2009) Hidden Markov Models are used for recognizing different maneuvers in terms of interactions between two traffic participants on extra urban roads. Thus, the temporal dependencies of the traffic scene are considered, and furthermore the simplified model structure is exploited. While in the previous version (Meyer-Delius et al., 2007) a discretization of the selected observations is done to handle the continuous input data (relative speed, velocity, acceleration between two traffic participants), in the newest version an approximation with different parametric probability distributions is applied. The estimation of the model parameters is done by using training data, while exact inference algorithms are used to execute the maneuver recognition. Testing is only performed with simulated data and with very few real world sequences consisting of two vehicles, what makes a statistical evaluation impossible.

Using HMMs may be a promising approach for recognizing traffic maneuvers, as been discussed in Chapter 2. Nevertheless, their usage in complex situations is still an open issue, as well as the evaluation of their performance. In real world situations with multiple objects (dynamic and static) the modeling of the traffic scene has to be adapted to keep the overall model complexity feasible.

Discussion: Different probabilistic approaches for the recognition of traffic maneuvers were developed in recent years, mainly using Bayesian Networks or their extensions to handle the spatio-temporal dependencies. These works are mostly focusing on basic urban scenarios (e.g., one specific intersection with given road geometry) or on simplified extra urban traffic situations (e.g. highways with only two traffic participants). In addition they have crucial difficulties to be evaluated with a significant large dataset of real world situations. Nevertheless, DBNs and especially HMMs have shown their potential to model difficult spatio-temporal dependencies and to be able to perform training and inference algorithms with reasonable computational cost. In the following, an approach for recognizing traffic maneuvers in extra urban scenarios using HMMs is presented in detail.

3.2. HMM-based Maneuver Recognition

In this section the basic idea of how to use HMMs for maneuver recognition is characterized and the different aspects of the modeling concept are depicted. Some exemplary results are outlined in Section 3.3.

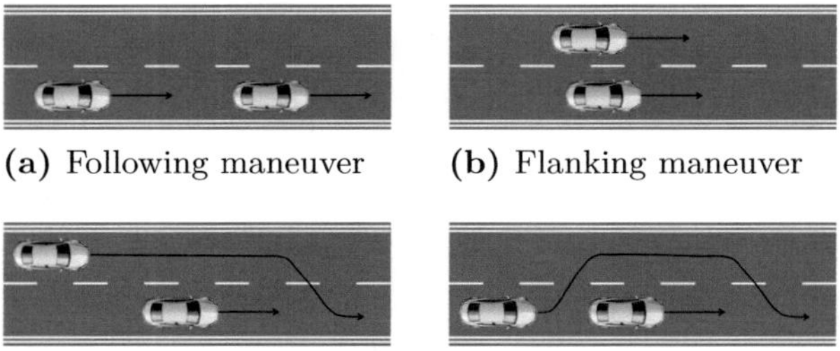

(a) Following maneuver

(b) Flanking maneuver

(c) Overtaking maneuver with 1 lane change

(d) Overtaking maneuver with 2 lane changes

Figure 3.2.: Paths of traffic maneuvers on extra urban roads.

3.2.1. Approach

The basic idea of this work is employing HMMs for modeling and recognizing traffic maneuvers as motivated by (Meyer-Delius et al., 2009) and published in (Firl and Tran, 2011) and (Firl et al., 2012). Typical extra urban maneuvers are *overtaking*, *following* and *flanking* maneuvers, which can be interpreted as interactions between different traffic participants. A schematic view is shown in Figure 3.2.

To recognize such maneuvers the following procedure is applied: Each maneuver is modeled by one HMM λ_i. For each point in time an observation o_t is extracted from the traffic scene (see Section 3.2.3) and for each model it is checked whether it is consistent with the current observation o_t (or with a sequence of observations $\mathbf{o}$) or not. Therefore the probability for each model given the respective observation sequence is calculated using the Forward algorithm[2]. The results are the probabilities $P(\mathbf{o}|\lambda_i)$ for each maneuver, whether it matches

[2] The Forward algorithm is used, because the concrete sequence of (hidden) system states is not of interest, but the probability of the entire model to be consistent with the observation. Otherwise Viterbi algorithm have to be used instead.

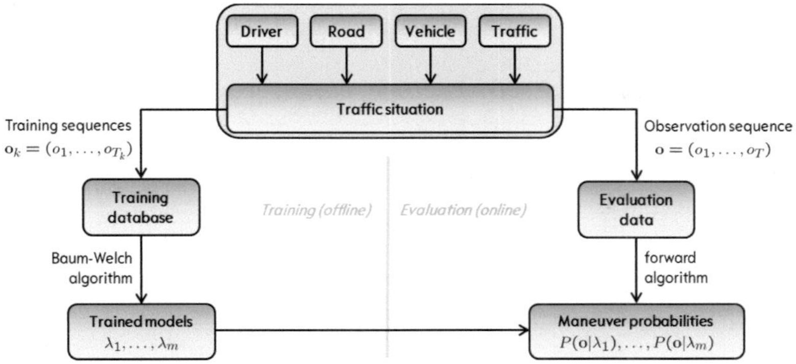

Figure 3.3.: Maneuver recognition approach including a (offline) training stage of m HMMs $\lambda_1, \ldots, \lambda_m$, and an (online) recognition stage.

the given observation or not. The decision which maneuver is currently been occurring is done by a comparison of all model probabilities. Hence, two different steps have to be done:

1. Training (Learning) of the model parameters
2. Recognition of traffic maneuvers

The training stage can be executed offline, since the model parameters are assumed to be constant over time and are thus only learned once. The recognition of traffic maneuvers must be performed online, since its results for an observation o_t is of interest at this particular point in time t. The approach is illustrated in Figure 3.3, where *Driver*, *Road*, *Vehicle* and *Traffic* are interpreted as input values for the traffic situation, which have to be considered by the observation $\mathbf{o}$. The training and recognition stage are both detailed in the following sections.

3.2.2. Training

Training in HMMs consists of two different tasks (see also Section 2.2.3): Determining the model structure and learning the model parameters $\lambda = (A, B, \pi)$. The first task is mainly related to defining the number of system states n for each model, restrictions to state transitions and the used observation model. After the structure has been determined, model parameters are estimated using the Baum-Welch algorithm.

Learning the model structure is done experimentally by using expert knowledge (see Section 2.2) and by keeping in mind the challenge of combining modeling capability and computational feasibility. Especially the number of system states n has to be chosen wisely, as the computational costs of performing one step of the Forward algorithm increases quadratically with the number of model parameters ($O(n^2T)$, see Section 2.2.3). However, increasing n always increases the modeling capabilities. Some criteria for model selection may be applied, as the Akaike information criterion (AIC) or the Bayesian information criterion (BIC), see (Zucchini and MacDonald, 2009) for more information. For real world applications model selection remains challenging as computational requirements and desired model capabilities are hard to determine. Different restrictions to state transitions $a_{i,j}$ defines different model structures, as *left-right models*, *input-output models* or *factorial HMMs*. In this work HMMs with 5 system states and no restrictions to the state transitions are used (fully connected HMM). This tourned out to be the most adequate tradeoff between computational effort and model capabilities. More system states would also require a larger training database to avoide the model only memorizing training data.

Besides the structure of the system states, the applied observation model plays a distinctive role for the complexity of the overall system. Since continuous observation values are usually acquired from the traffic scenario (see Section 3.2.3), a discretization or parametrized probability functions have to be used for approximation (see Section 2.2). To avoid the shortcomings of a discretization of the range of the observation values, a Mixture of Gaussians model (MoG) is applied to approximate the exact observation distribution $P(O_t|X_t)$:

$$b_i(o) = P(O_t = o|X_t = x_i) \approx \sum_{j=1}^{n_{mix}} c_{i,j} \mathrm{N}(o, \mu_{i,j}, \sigma_{i,j}^2), \tag{3.1}$$

where n_{mix} is the number of used mixtures and $c_{i,j}$ are the weighting coefficients. For applications, with more than one observation available, i.e., where o consists of multiple observation values, Eq. (3.1) has to be considered for all observations o. In the following, $\mathbf{o}$ and o_t denotes the complete vector of n_{obs} observations rather than single values.

For learning all model parameters the Baum-Welch algorithm is applied for every maneuver to estimate the state transition matrix $A \in \mathbb{R}^{5\times5}$, the initial state distribution $\pi \in \mathbb{R}^5$ and the observation model B. The parameters of the observation model are c, μ and σ for every combination of possible system state and observation and for every Gaussian mixture. The number n_{mix} is chosen according to the number of system states and the available hardware performance (in this work $n_{mix} = 3$). The precise definition of the observation vector is presented in the following Section 3.2.3.

The training database is generated using the simulation tool CarMaker® (IPG Automotive GmbH, 2011). Real world data was not used for training purposes, because of the great number of required training sequences due to the large number of model parameters and due to the variety of how maneuvers can be performed[3]. In this work the number of parameters for each model is

[3] A crucial difficulty when using real world data for training purposes is that for an entire setup of the database, safety critical maneuvers have to be considered, especially for dynamic overtaking maneuvers.

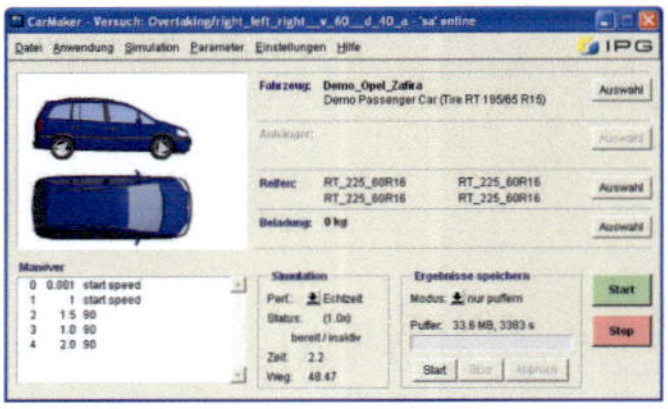

(a) CarMaker main frame.

(b) IPG Movie frame for visualization.

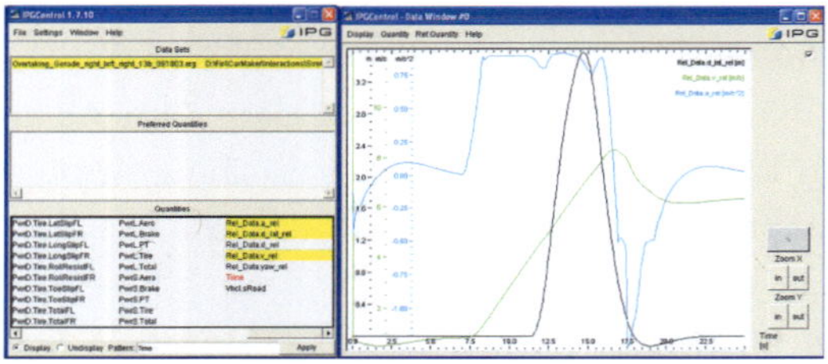

(c) IPG Control for simulation analysis.

Figure 3.4.: IPG CarMaker®simulation environment

$$\underbrace{5*5}_{A}+\underbrace{5}_{\pi}+\underbrace{5*3}_{c}+\underbrace{4*5*3}_{\mu}+\underbrace{4*4*5*3}_{\sigma}=345, \tag{3.2}$$

where c, μ and σ are the MoG parameters of the observation model B. CarMaker® is able to simulate different driver, road and vehicle characteristics as well as multiple traffic participants. An analysis tool provides a fast evaluation of the generated simulation sequences, depicted in Figure 3.4. For training the used models about 900 sequences of overtaking maneuvers, 100 sequences of following maneuvers and 50 sequences of flanking maneuvers are acquired. Their relative paths can be seen in Figure 3.5. The definition of the coordinate system and the complete observation vector is given in the following Section 3.2.3. When creating the databases, different parameters are identified to be crucial for the execution of the maneuver and have to be varied throughout the generation process, e.g.:

- different driver characteristics, i.e., aggressive $\leftrightarrow$ defensive drivers, including different acceleration values,
- different absolute velocities of the vehicles,
- different lane positioning of the vehicles (relative to center line),
- different distances between the vehicles (longitudinal and lateral) when performing the maneuver,
- different relations of distances and relative and absolute velocities of the vehicles.

Without loss of generality all sequences are simulated on straight roads, see Section 3.2.3, and with two vehicles performing the maneuver, see Section 3.2.4. After an initialization Baum-Welch algorithm estimates the model parameters with this data resulting in (locally) optimal model parameters. As no global optimality is guaranteed, initial parameters have to be chosen wisely and with respect to the used training data. Furthermore, the training procedure has to be applied multiple times and with different initializations to minimize the risk of selecting non-optimal model parameters.

3.2.3. Observation Model

For the complete maneuver recognition approach the definition of the observation vector is a key aspect, as it defines the scenario information of the traffic scene which is considered. The more accurate the observation vector is defined, the more robust the recognition framework will work. Keeping in mind that the approach presented in this work should be handle different sensor setups working as input data source, no special information about the traffic scene, which can be observed hardly

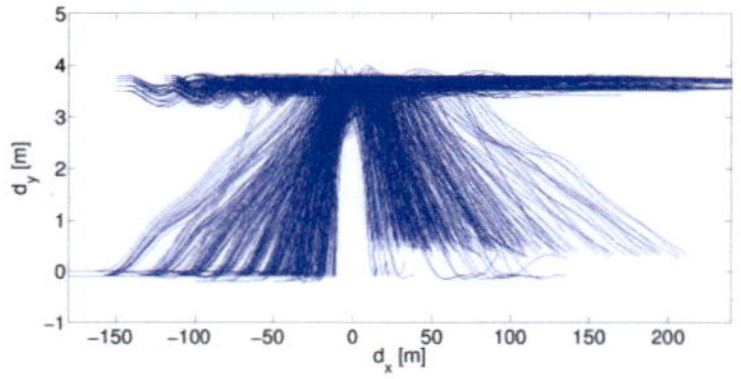

(a) Database for overtaking maneuver.

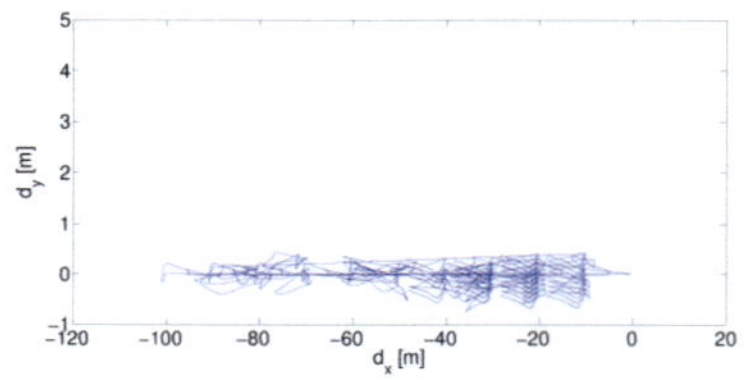

(b) Database for following maneuver.

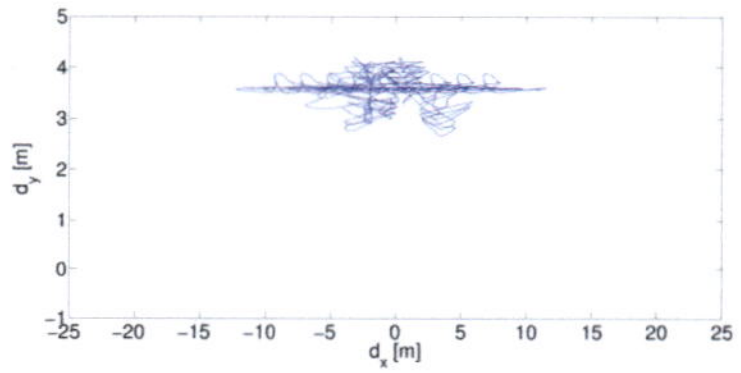

(c) Database for flanking maneuver.

Figure 3.5.: Relative paths of sequences of simulated database for overtaking, following and flanking maneuvers.

with common sensor setups, have to be used. Examples for this are:

- blinker (direction indicator),
- information on the driver (viewing direction, driver's mood),
- information of infotainment system (navigation).

Thus, the observation vector mainly consists of the kinematic relation of the two vehicles performing the maneuver, i.e.:

$$o_t := (d_x, d_y, v_{rel}, a_{rel}), \tag{3.3}$$

where d_x and d_y are the relative distances in two dimensions and v_{rel} and a_{rel} are the relative velocity and acceleration,

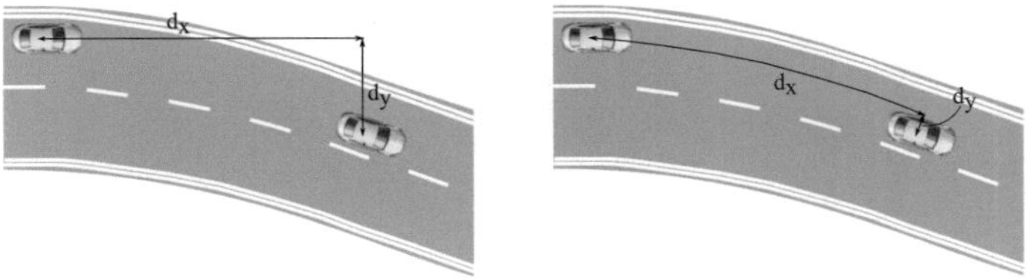

(a) Vehicle fixed Cartesian coordinate system.

(b) Lane fixed coordinate system.

Figure 3.6.: Motivation for the use of a lane fixed coordinate system on a right-hand bend with the observation d_x and d_y.

respectively[4]. For application purposes the underlying coordinate system has to be defined, since different types are possible and feasible.

In most cases observations are received by a vehicle's on-board sensor system, like radar or camera systems. The choice of using a sensor-fixed coordinate system is therefore often a straightforward approach (a vehicle-fixed coordinate system can be used equivalently in many cases). While on straight roads this usually results in reasonable observations, the road's geometry has to be taken into account on winding streets. In Figure 3.6 the undesired effect for the observations d_x and d_y is depicted, when using a vehicle fixed coordinate system at a right turn. Especially the value of d_y in Figure 3.6(a) is not characteristic for the present scene when not taking the road geometry into account. The resulting paths for both systems, when two vehicles perform an overtaking maneuver at a left curve can be seen in Figure 3.7. While the paths in the lane fixed coordinate system can be clearly interpreted, the vehicle

[4] v_{rel} and a_{rel} are only used as 1-dimensional values, because most common automotive sensors have major difficulties in providing accurate resolutions of the velocity and acceleration of observed objects in longitudinal and lateral direction.

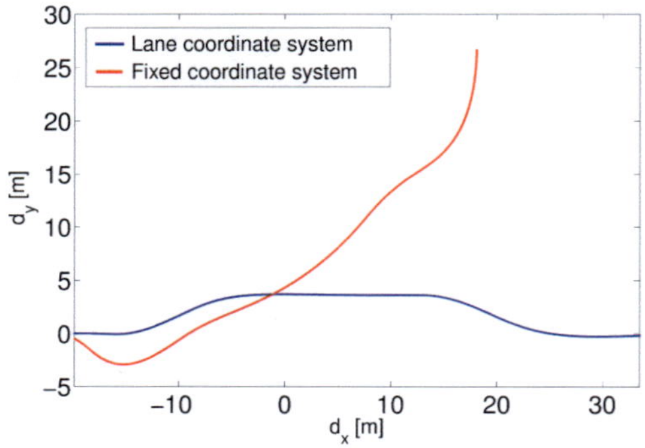

Figure 3.7.: Paths of an overtaking maneuver with two different coordinate systems at a left curve. Lane adapted coordinate system and fixed coordinate system of the overtaken vehicle (coordinates of maneuver start: (−20|0)).

fixed coordinate system results in less obvious paths. The consequences of that on the maneuver recognition results will be discussed in Section 3.3.

To transform the observation data of the sensor fixed coordinate system into the lane oriented one, a geometrical model of the lane has to be acquired. This depends mainly on the used sensor setup, and thus cannot be assumed to be given in all situations[5]. Two commonly used models for representing the geometry of the lane are splines (e.g., presented in (Hasberg and Hensel, 2008) and (Hasberg et al., 2012)) and clothoids (introduced in (Archibald, 1918)), which are depicted in Figure 3.8. These are used in the experiments presented in Section 3.3. Clothoids are planar curves with a curvature changing linearly with the curve length L. Their parameters are the lateral offset Y_E, the offset angle Φ and the curvature parameters c_0 and c_1.

[5]If no information of the geometry of the lane can be extracted out of the sensor data, the sensor or vehicle coordinate system has to be used instead.

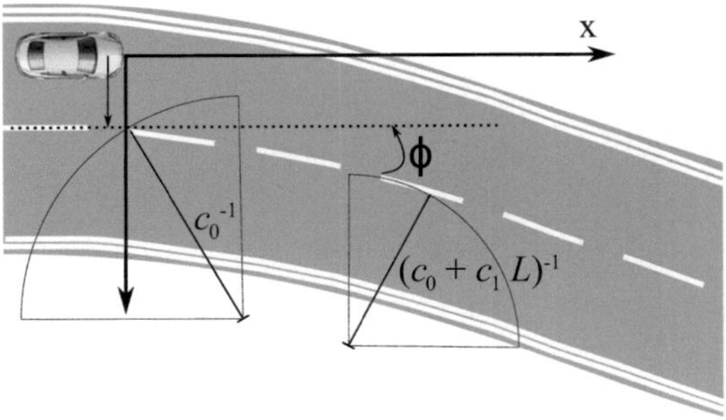

Figure 3.8.: Clothoid model with length L, parametrized by (Y_E, Φ, c_0, c_1); see (Darms et al., 2010)

Since all entries of the observation vector in Eq. (3.3) are usually continuous valued, an approximation using Mixture of Gaussians have to be applied for all of them. Their entire parameter set (c, μ, σ) are trained using Baum-Welch algorithm with $n_{mix} = 3$. In Figure 3.9 the resulting probability distribution for the observations d_x and d_y are plotted for all 5 system states of the *following* model. For the recognition of traffic maneuvers, the distributions of single system states x_i does not play a significant role, but the complete distributions over all states does, as depicted in Figure 3.9(f) (see Algorithm 1, summation in the last line).

3.2.4. Free Space Consideration

The observation vector from Eq. (3.3) considers the relational kinematics between two traffic participants. However, this restricts the modeling capability in many real world situations, where more than two vehicles have to be taken into account, or other static objects influence the execution of traffic maneuvers. To compensate this drawback one possible solution is to extend the observation vector with the relative kinematic information

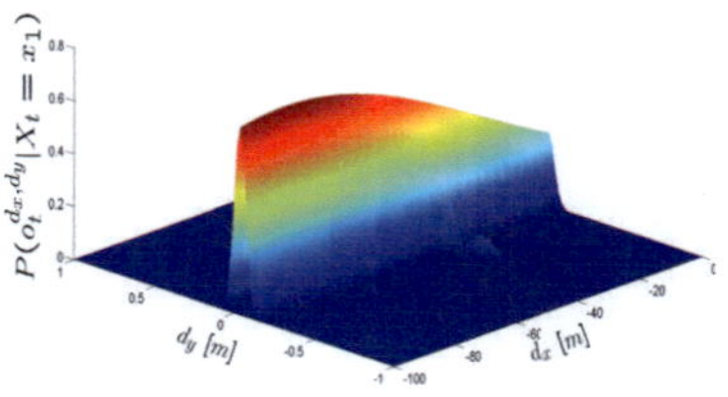

(a) State $X_t = x_1$

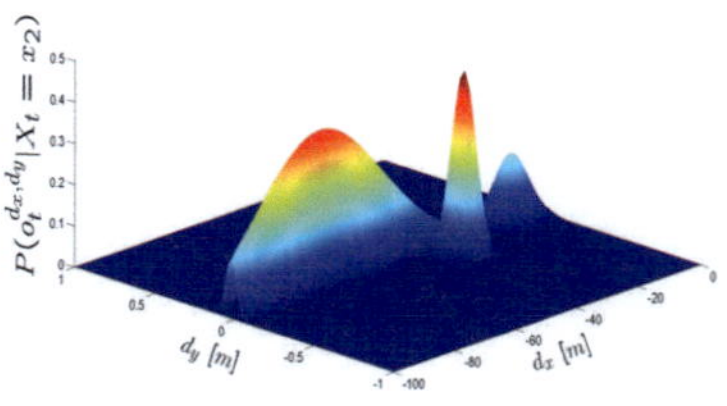

(b) State $X_t = x_2$

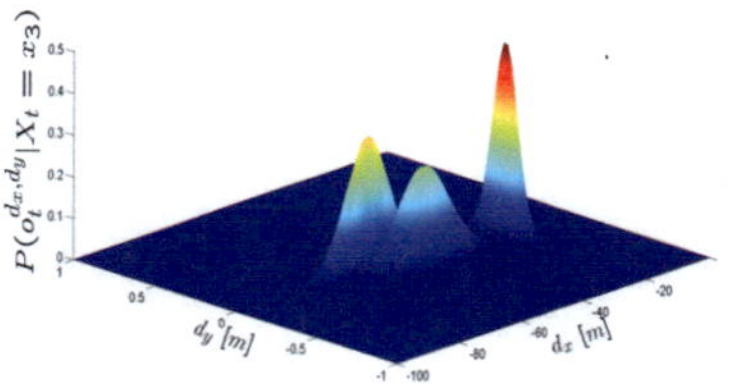

(c) State $X_t = x_3$

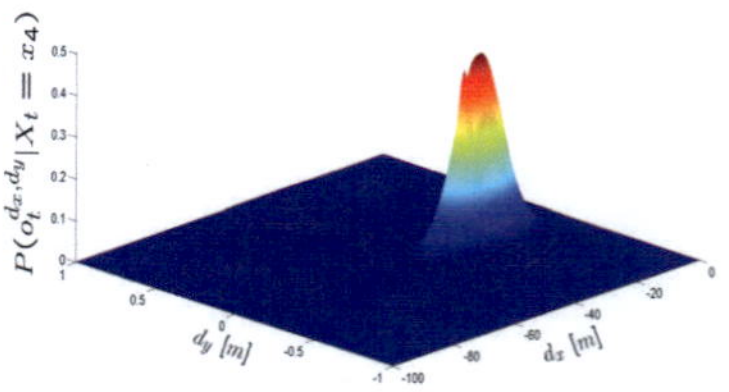

(d) State $X_t = x_4$

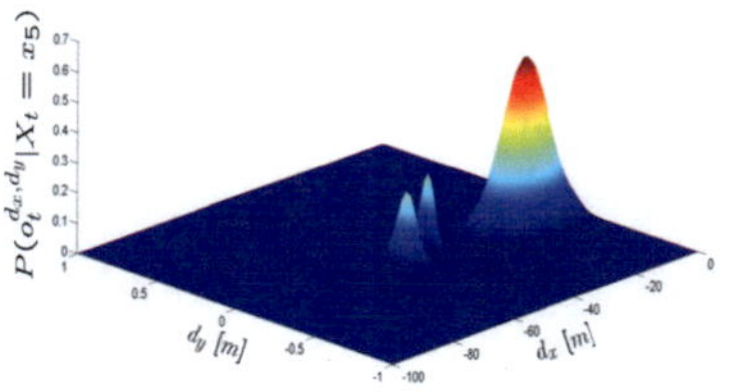

(e) State $X_t = x_5$

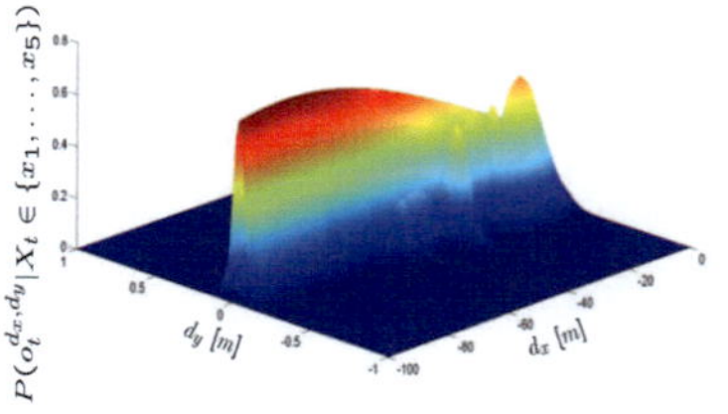

(f) State $X_t \in \{x_1, \ldots, x_5\}$

Figure 3.9.: Observation distributions for the 2 observations d_x and d_y of the following model for different given system states X_t and their weighted sum ($n_{mix} = 3$). The different system states can be interpreted as different characteristics of following maneuvers, i.e., mostly different longitudinal distances.

to all others vehicles, e.g., distance, velocity and acceleration data to the next vehicle ahead (and/or behind) in the same lane, or in all adjacent lanes. This approach increases the dimensionality of the observation vector significantly and makes it unfeasible for most applications. In this work a method based on occupancy states is presented that only increases the

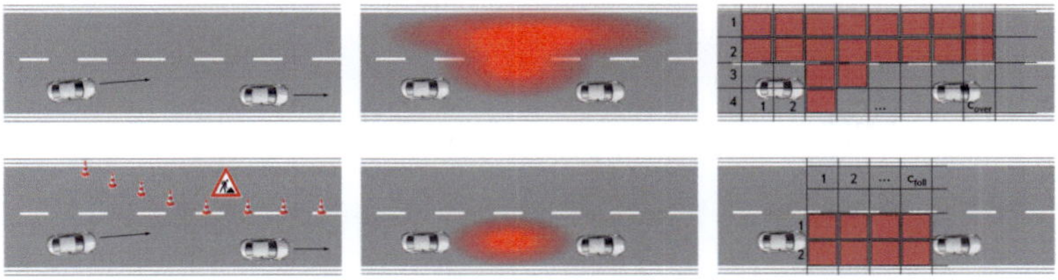

(a) Motivation for free space consideration (example: overtaking recognition).

(b) Required free space for overtaking and following maneuvers.

(c) Definition of the occupancy grid for overtaking and following maneuvers.

Figure 3.10.: Usage of occupancy grids for free space consideration, see (Firl et al., 2012).

dimensionality of the observation vector by 1 (see also (Firl et al., 2012) and (Firl and Tran, 2011)). Nevertheless, it is possible to consider a wide range of influences on the traffic scene:

- other traffic participants and their future driving trajectories,
- static objects, like construction sites or road boundaries,
- road characteristics, like different number of lanes.

An illustrative example is shown in Figure 3.10: In both situations of Figure 3.10(a) the same observation vector $\mathbf{o}$ of Eq. (3.3) is extracted. This implies the same maneuver probabilities $P(\mathbf{o}|\lambda_i)$ for all maneuvers λ_i (see Figure 3.3). Obviously the probability that the two vehicles performs, let's say, an overtaking maneuver is definitively different. In the situation of the lower image of Figure 3.10(a), an overtaking maneuver is even impossible due to the construction site on the left lane. So, the obtained observation model does not reflect the situation appropriately.

The main idea of considering such influences is to model the required free space for the execution of the maneuver and to add this information to the observation vector. This is done by the following steps (published in (Firl et al., 2013)):

First, the required free space for a maneuver has to be defined using human expert knowledge, see Figure 3.10(b) for examples of *overtaking* and *following* maneuvers. As the interacting traffic participants changes their positions relative to each other, the dimensions of the available free space may vary over time.

The next step is to consider the information of this required free space in a mathematical way, which can be either *free*, *occupied* or neither complete free nor complete occupied, which is detailed later. For this purpose occupancy grids are used in this work as depicted in Figure 3.10(c). The number of cells varies for different maneuvers, but they are fixed for one given maneuver. Furthermore, the number of cells is independent of the relative distances of the interacting vehicles and thus, the size of each cell is variable. The dimensions of the grid depends on the accuracy of the given sensor information.

To combine this information of the occupancy grid with the given observation model, each cell of the grid has to be adapted with a value representing either the cell is free for the execution of the maneuver, or it is blocked (or neither). Therefore the variables $f_{i,j} \in [0,1]$ are defined for every maneuver λ as follows:

- $f_{i,j} = 0$: the cell (i,j) is occupied and restricts an execution of the maneuver λ.
- $f_{i,j} = 1$: the cell (i,j) is free and no restrictions to the execution of the maneuver λ are given.
- $f_{i,j} \in (0,1)$: the cell (i,j) is neither completely free nor completely occupied, and thus the execution of the maneuver is neither definitively possible nor impossible (regarding this cell).

While the first two cases ($f_{i,j} \in \{0,1\}$) mainly represent the consideration of static objects (like construction sides or road boundaries), other values correspond to the consideration of other dynamic traffic participants. Because their relative distances change over time, not only the current position has to be taken into account, but also their future driving trajectory. Because there exist a lot of different prediction approaches, and the prediction accuracy is highly depending on the used sensor data, a general reasonable formula for the values $f_{i,j}$ is hardly possible. Nevertheless, the interpretation is: *The more accurate the prediction regarding the occupancy of a cell* (i,j) *is, the closer the value* $(f_{i,j})$ *is to* 0.

As a last step, the observation vector o_t has to be adapted by adding a new, continuous observation value $f \in [0,1]$ resulting in

$$o_t := (d_x, d_y, v_{rel}, a_{rel}, f). \tag{3.4}$$

This adapted observation model has to be considered in the training data for each model. f is calculated for the maneuver recognition stage (and also for the training data) by summing up over all cells:[6]

$$f = \frac{1}{\#cells} \sum_{\text{all cells } (i,j)} f_{i,j}. \tag{3.5}$$

The proposed approach of using adaptive occupancy grids for modeling different characteristics of traffic scenarios combines different advantages. It allows to consider all possible objects having any influence of the execution of the traffic maneuver. This also implies multiple, dynamic traffic participants of a varying number. At the same time, the dimension of the obser-

[6] A weighted sum over all cells is also possible, but then the single weights for each cell have to be trained manually by using human expert knowledge.

vation model is only increased by 1 keeping the computational complexity at a low level.

3.2.5. A Priori Knowledge

The described observation model (Section 3.2.3 and Section 3.2.4) is able to capture most relevant information for recognizing traffic maneuvers. However, it still lacks in taking different road types and different driver characteristics (aggressive vs. defensive) into account. Before presenting a possibility to add this information to the overall recognition approach, the following characteristic of the recognition results have to be noticed: The calculated probability $P(\mathbf{o}|\lambda_i)$ of an observation sequence $\mathbf{o}$ given the model λ_i is not the crucial quantity for decision making, but the a posteriori probability $P(\lambda_i|\mathbf{o})$ is ('given' is not the model, but the observation sequence). The probability of an observation sequence $\mathbf{o}$ is not of interest, but the probability of the model λ_i.

Therefore, the Bayes' theorem is applied as follows, assuming that the observation $\mathbf{o}$ is independent of the model λ_i and is thus assumed to be constant over all models, and $P(\mathbf{o}) \neq 0$ holds:

$$P(\lambda_i|\mathbf{o}) = \frac{P(\mathbf{o}|\lambda_i) \cdot P(\lambda_i)}{P(\mathbf{o})} \propto P(\mathbf{o}|\lambda_i) \cdot P(\lambda_i). \qquad (3.6)$$

Since the probability of the observation sequence $\mathbf{o}$ is constant for all models λ_i, $P(\mathbf{o}|\lambda_i))$ may also be used for recognizing traffic maneuvers. The a priori probability $P(\lambda_i)$ may now be used to model different characteristics of the traffic scenario (see also (Firl and Tran, 2011)):

- *Different types of road:* An example is, that following maneuvers are generally more likely to occur on single-lane roads than on multi-lane roads and thus, the corresponding probability $P(\lambda_{foll})$ has to be larger when driving on single-lane roads.

- *Different driver characteristics:* An example is, that aggressive, fast driving people are generally more likely to execute overtaking maneuvers than defensive ones and thus $P(\lambda_{over})$ has to be larger when identifying aggressive drivers.

Both types of information are not available in every situation and with every sensor setup. Therefore, in these cases the probabilities $P(\lambda_i)$ are assumed to be uniformly distributed over all models λ_i. Information on driver characteristics may be gathered either from recognized driving maneuvers of the past, or from the current driving mode (e.g., sportive, touring) of the ego vehicle. The probabilities of different type of roads have to be calculated by simple counting different maneuver types over a large set of sequence on the specific road type. Furthermore the current type of road have to be known using digital maps.

3.3. Maneuver Recognition Results

After describing all relevant input for the maneuver recognition in the last section, in the following some exemplary results of the HMM-based approach are presented. As depicted in Figure 3.3 the trained models λ_i are evaluated with respect to the current observation sequence $\mathbf{o}$ using the *forward* algorithm.

Before implementing the proposed algorithmic approach, one property of the forward variables $\alpha_t(i)$ have to be considered: Their computation requires a sum over many numbers, which tend exponentially to zero each (see (Rabiner, 1989) for more information: $\alpha_t(i)$ consists of terms which can be written as $\prod_{s=1}^{t-1} a_{x_s x_{s+1}} \prod_{s=1}^{t} b_{x_s}(o_s)$, where each value of a and b is significantly smaller than 1). To avoid an exceeding of the precision range of most hardware platforms especially for large observation sequences $\mathbf{o}$, the following scaling procedure is applied in this work, according to (Levinson et al., 1983). The scaling factor c_t is defined as

$$c_t := \left[\sum_{i=1}^{N} \alpha_t(i) \right]^{-1}, \tag{3.7}$$

independently of the state X_i, only dependent on the point in time t. It is applied to $\alpha_t(i)$ resulting in:

$$\hat{\alpha}_t(i) = \frac{\alpha_t(i)}{\sum\limits_{i=1}^{N} \alpha_t(i)} = c_t \alpha_t(i), \tag{3.8}$$

with $\sum_{i=1}^{N} \hat{\alpha}_t(i) = 1$.

Simulated Data: A first verification of the reliability of the proposed concept for recognizing traffic maneuvers is realized with simplified sequences $\mathbf{o}_i$ of simulated data not contained in the training database: $\mathbf{o}_i \notin \mathbf{o}_{train}$. Therefore, 10 evaluation sequences for each of 3 different types of overtaking maneuvers are generated. The recognition results for λ_{over} and λ_{foll} are normalized in their length using a time warping algorithm as can be seen in Figure 3.11. Furthermore, the a priori probabilities $P(\lambda_i)$ are assumed to be equal for all models.

The results can be correctly interpreted for all 3 different situations. For the overtaking maneuvers with no lane change, a distinction between the overtaking and following maneuver may be easily performed for all points in time. For both other situations, the two models are more difficult to distinguish as both vehicles start the maneuver on the same lane. In these situations a following maneuver may not be excluded at the beginning of the sequence. The verification of sequences contain-

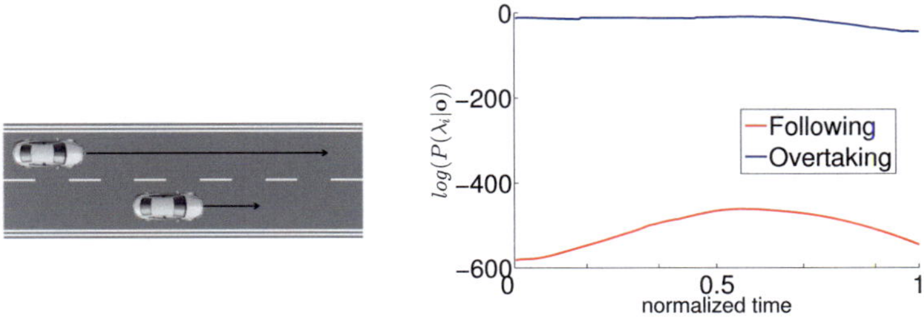

(a) Overtaking maneuvers with no lane change.

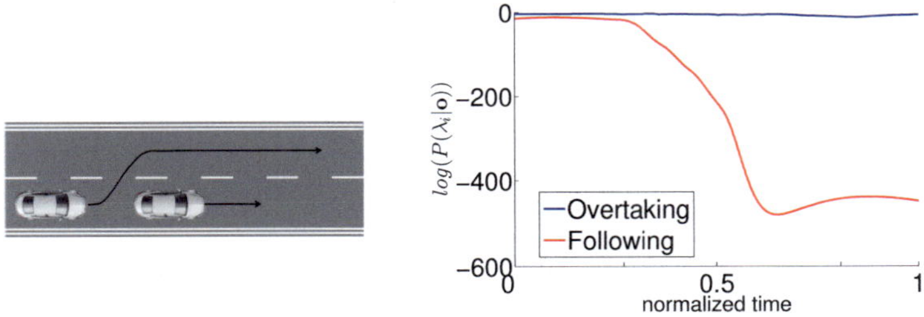

(b) Overtaking maneuvers with 1 lane change.

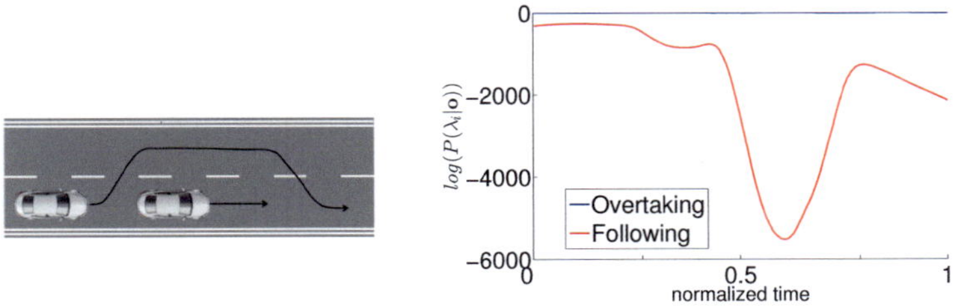

(c) Overtaking maneuvers with 2 lane changes.

Figure 3.11.: Simulated Results of the maneuver recognition concept for the models *following* and *overtaking*: Left: paths of the maneuvers; Right: normalized results over 10 similar sequences each.

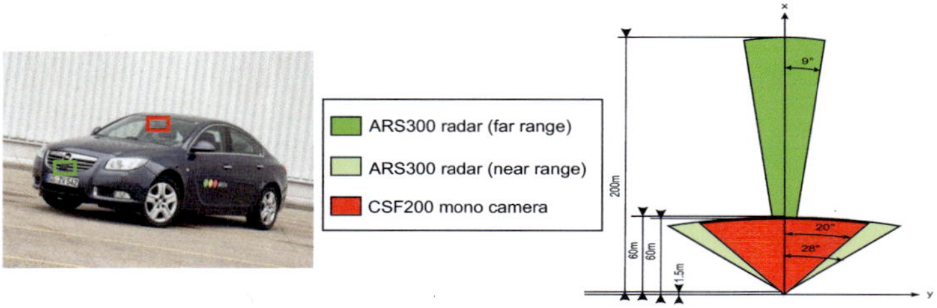

Figure 3.12.: Sensor setup of the used experimental vehicle: Sensor system consists of a mono camera and a radar sensor performing a far and a short range scan. Left: Opel Insignia experimental vehicle. Right: Opening angles of the used sensor setup (Firl and Tran, 2011).

ing other maneuvers, or even more complex situations implying multiple stages of different maneuvers, shows results which are interpretable in a similar way.

Real World Data: For testing the proposed framework in real world scenarios, data was recorded with the sensor setup illustrated in Figure 3.12. It consists of a mono camera (Continental CSF200) and a radar sensor (Continental ARS300), performing near- and far range scans. The underlying fusion concept provides information about all surrounding objects with 25 frames per second. The vehicle is used as an observer, to gather information about vehicles performing different maneuvers in front of the car. While the radar sensor is mainly used for acquiring relative kinematic information (position, speed, acceleration), the camera is employed for approving objects in the sensor fusion component and for visual verification purposes. More information about the sensor setup is given in Appendix A.1. A sequence of 2 vehicles performing an overtaking maneuver on a two-lane road can be seen exemplary in Figure 3.13.

Exemplary results of the sequence containing the screenshots of Figure 3.13 are depicted in Figure 3.14. In this example two vehicles are driving on a two-lane road: At the first

Figure 3.13.: Screenshots of one testing sequence with two vehicles performing an overtaking maneuver on an extra urban street (Firl and Tran, 2011). Maneuver recognition results are shown in Figure 3.14. Screenshots are captured approximately at frame 0, 270 and 560.

stage, the vehicles are following each other. After that the rear vehicle accelerates and suddenly overtakes the other one performing two lane changes. The first lane change is executed approximately at frame 250, the second lane change at frame 520. The three different stages of the sequence of Figure 3.13 can be clearly separated as the results are easy to interpret. The probabilities $P(\lambda_i)$ according to Section 3.2.5 are set manually representing a two-lane road in extra urban scenarios, i.e., $P(\lambda_{following}) = 0.9$ and $P(\lambda_{overtaking}) = P(\lambda_{flanking}) = 0.05$. Because the used radar sensor does not provide accurate resolution for the lateral position and velocity, the noisy maneuver probability $P(\lambda_{flank}|\mathbf{o})$ for *flanking* maneuvers becomes explicable. However, the lateral relative position d_y is a crucial information for recognizing flanking maneuvers.

Similar results are achieved with similar evaluation sequences with the same sensor setup, which provide the reliability of the proposed approach, since recognition results matches with human expert perception of the traffic scene. Nevertheless, a

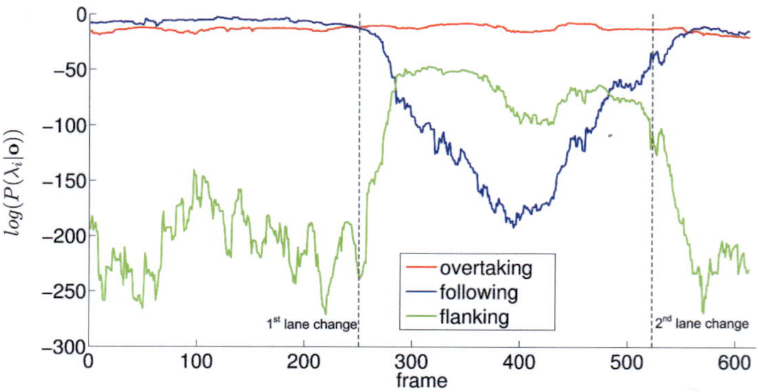

Figure 3.14.: Results of the maneuver recognition approach of a testing sequence of two vehicles on a extra urban road for the overtaking, following and flanking model. Results are plotted logarithmically due to very small probability values of the forward algorithm. For evaluating both following maneuvers simultaneously (vehicle 1 follows vehicle 2 and vehicle 2 follows vehicle 1), the absolute longitudinal distance $|d_x|$ is used.

mathematical evaluation is not possible, due to the missing of ground truth data, which may be interpreted differently be human experts. Thus, the presented probability curves only allow a visual verification of the proposed approach. Possible evaluation methods will be presented in Chapter 4.

Coordinate Systems: To motivate the usage of a lane fixed coordinate system as introduced in Section 3.2.3, the results of the simulated sequence of Figure 3.7 with two different coordinate systems are depicted in Figure 3.15. The maneuver probabilities using the Cartesian coordinate of Figure 3.15(b) are only plausible in the first few frames. After that, the results become less comprehensible, e.g., the increasing probability of the following model approximately at frame 100. Furthermore, all probability values are decreasing very fast, so that they exceed the precision range sooner or later. Therefore, only short sequences may be evaluated in that way.

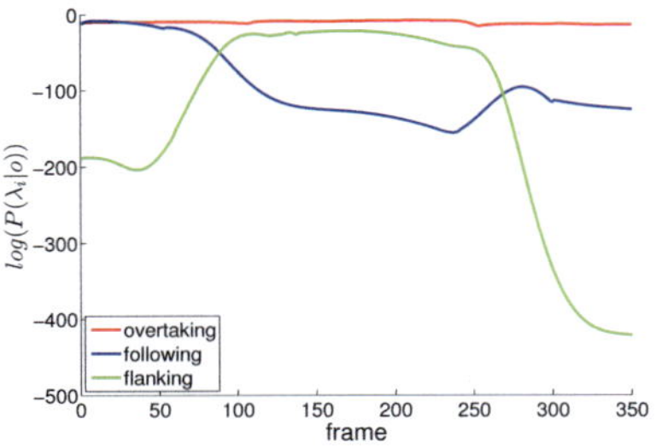

(a) Lane oriented coordinate system

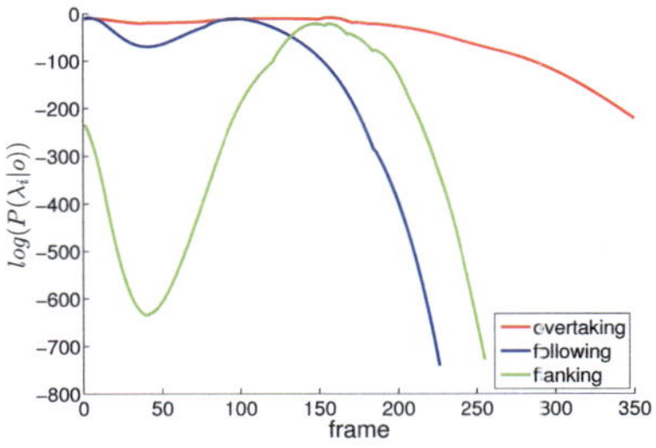

(b) Fixed (Cartesian) coordinate system

Figure 3.15.: Results of Maneuver Recognition with two different coordinate systems of a simulated overtaking maneuver with two lane changes. While the results with the lane oriented coordinate system can be easily verified, the results with the world fixed, Cartesian coordinate system cannot.

3.4. Summary

In this chapter a framework for recognizing maneuvers in extra urban traffic situations was presented. After reviewing some state of the art recognition approaches, a probabilistic framework based on Hidden Markov Models was introduced. To consider most influences of real world scenarios, different parameters are employed in the observation model.

Primary observation input is the relative kinematic information between two interacting traffic participants (observation vector Eq. (3.3)): the longitudinal and lateral relative distance, relative velocity and acceleration values.

To capture more complex situations, adaptive occupancy grids were introduced to model the feasibility of a maneuver execution. Their occupancy values are integrated in the observation vector resulting in Eq. (3.4). Thus, the following aspects of real world traffic situations are taken into account:

- dynamic traffic participants and the influence of their predicted trajectories on the traffic scene,
- road characteristics, e.g., boundaries, construction sides or other static objects.

A priori probabilities are integrated using Bayes' theorem to consider the dependencies of single maneuvers on different road types or on different driver characteristics.

Recognition was performed for the defined observation vector using the *forward* algorithm and the resulting probabilities for each model are compared to each other. Recognition results were presented for exemplary sequences of simulated as well as for real world data. The general reliability of the proposed approach was shown as recognition results match well with human expert perception of the corresponding traffic scene.

4. Evaluation and Application of Maneuver Recognition

After presenting the maneuver recognition approach in the last chapter, the focus of the following sections is twofold:

- How may the proposed concept be evaluated?
- What is the possible benefit of the proposed concept for a specific ADAS feature?

The first question about the evaluation of the framework of Chapter 3 cannot be answered in a general way. Evaluation of all works in the field of traffic situation assessment deals with the problem that it highly depends on the triggered functions. That means it depends on the proceeding steps of the ADAS information flow, depicted in Figure 1.1.

In Section 4.1 a method is presented to evaluate the recognition concept in terms of the prediction of single, safety critical traffic maneuvers. They are crucial for most state of the art ADAS. A possible application area therefore is outlined afterwards in Section 4.2, where the presented maneuver prediction is applied to Car-to-X (C2X) communication and its security issues.

4.1. Single Maneuver Prediction

The evaluation of concepts for situation recognition depends on the ADAS function it is used for, i.e. it depends on the *behavior generation* stage (Figure 1.1). This implies that it also depends on the specific situations the ADAS feature is designed for. In this work an evaluation is proposed which takes traffic situations into account that are commonly addressed by most systems to be safety critical.

Typical critical traffic situations, where an ADAS may execute any intervention or driver warning strategy, are usually characterized by suddenly occurring changes of the movement of any traffic participant. These can be grouped into

- lateral maneuvers, mainly lane changes,
- longitudinal maneuvers, mainly abrupt brakings,

where only extra urban scenarios are considered. For the best capability of a given assistance function, it is crucial to recognize or predict these situations as early and as robustly as possible. However, an evaluation for both aspects (prediction time and robustness) is not available in general as their importance may differ from one feature to another. Moreover, there are opposing requirements as an early prediction will usually degrade robustness and vice versa.

Hidden Markov Models are able to handle the spatial dependencies of the modeled situation as well as its temporal components. Therefore, the HMM based model approach may be evaluated due to their prediction capabilities for the situations mentioned above.

A first approach may address the exact prediction of possible vehicle paths (or even trajectories) from the calculated maneuver probabilities $P(\lambda_i|\mathbf{o})$. Using the trained observation models, i.e. the probability distributions $P(O_t = o_t|X_t = x_i)$ of the observation o_t at time t while being in state x_i, enables the usage of sampling techniques for prediction purposes. Unfortunately, the used modeling concept does not use any kinematic motion model, so changes in system states do not correspond to exact changes of the vehicles physical states. A direct consequence is that using the observation probability distributions

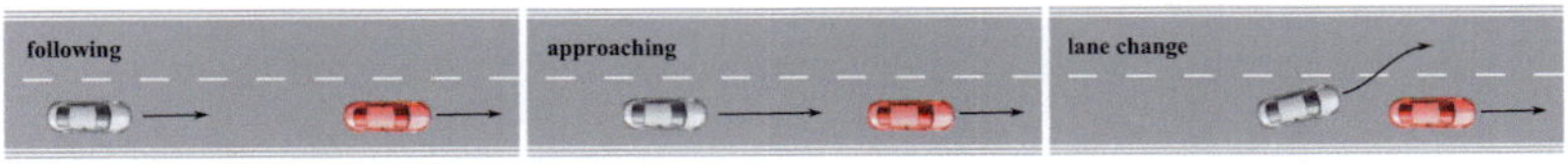

Figure 4.1.: Typical flow of a lane change maneuver on extra urban traffic roads. The three stages are: following (optional), approaching and lane change; see (Firl et al., 2012).

for predicting vehcile's trajectories may result in non-plausible vehicle states. For example, possibly predicted states of the observation vector o_t of Eq. (3.4) may include positions d_x, d_y, which does not correspond to prevision positions and velocity values v_{rel} with respect to the applied time base. A method for predicting future vehicle's trajectories with respect to the calculated maneuver probabilities is presented in Chapter 5.

In the following sections the maneuver probabilities are directly used for predicting single traffic maneuvers. The absolute probability for a given maneuver at a specific point in time does not allow any concrete interpretation, since it highly depends on the used training data and the contained sensor noise. Therefore, a comparison of different model probabilities is used instead. Thus, the ratio of two models λ_1 and λ_2 and a given observation sequence $\mathbf{o}$ is evaluated by

$$c = \frac{P(\lambda_1|\mathbf{o})}{P(\lambda_2|\mathbf{o})} = \frac{P(\mathbf{o}|\lambda_1)P(\lambda_1)}{P(\mathbf{o}|\lambda_2)P(\lambda_2)}. \tag{4.1}$$

Evaluation is done afterwards with respect to different ratios of correctly predicted and not correctly predicted situations (true and false positives).

4.1.1. Lane Change Prediction

A typical lane change maneuver on an extra urban road is illustrated in Figure 4.1 as it is usually performed during overtaking maneuvers. It can be roughly divided into the following stages:

1. *following*: Both vehicles are driving in the same lane with approximately the same speed. This stage is sometimes omitted, but may occur at the end of speed limited zones, as an example.

2. *approaching*: The rear vehicle is approaching. This stage varies much in its length depending mostly on specific driver characteristics.

3. *lane change*: The rear vehicle is performing the lane change in order to overtake.

Prediction of the lane change is done by using the ratio of Eq. (4.1) with the models for overtaking λ_{over} and following λ_{foll}:

$$c_1 = \frac{P(\lambda_{over}|\mathbf{o})}{P(\lambda_{foll}|\mathbf{o})}. \tag{4.2}$$

Large values of c_1 thereby indicate high lane change probabilities. Evaluation of this approach is done with respect to the point in time of prediction which depends on c_1. A lane change should be usually predicted the sooner the better but at the latest before stage 3 starts, i.e. before the real lane change is executed. Crucial values for the evaluation purpose are the number of situations with correctly predicted lane changes (true positives) and the number of situations where a lane change is incorrectly predicted (false positives).

4.1.2. Abrupt Braking Prediction

A typical abrupt braking maneuver on an extra urban road is illustrated in Figure 4.2. The reason for the abrupt braking maneuver is a blocked lane on the left because of a passing vehicle. Of course, other reasons for the blockage of the lane are also possible, as already mentioned in Section 3.2.4. These maneuvers can be roughly divided into the following stages:

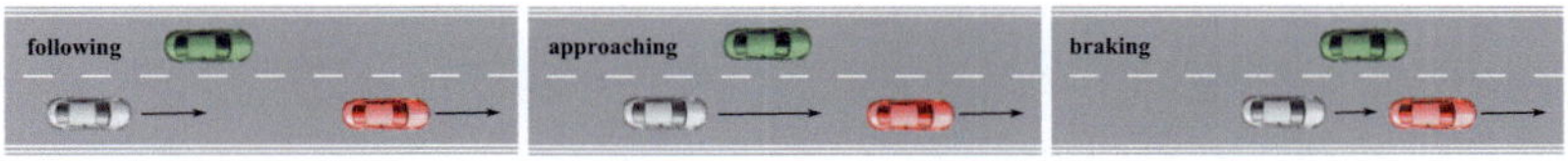

Figure 4.2.: Typical flow of an abrupt braking maneuver on extra urban traffic roads. The three stages are: following (optional), approaching and braking; see (Firl et al., 2012).

1. *following*: Both vehicles are driving in the same lane with approximately the same speed. This stage is sometimes omitted.

2. *approaching*: The rear vehicle is approaching, motivated by the demand on an overtaking maneuver.[1] This stage varies much in its length depending mostly on specific driver characteristics.

3. *braking*: The rear vehicles have to execute a braking maneuver because of the blockage of the left lane.

Predicting these situations accurately requires a slightly different approach than used for lane change maneuvers. The main reason for abrupt brakings of Figure 4.2 can be interpreted as a general divergence of the driver's intention to overtake another vehicle and the feasibility of the maneuver. Therefore, the following two models are acquired:

- The *driver's intention* to overtake another vehicle is represented by the model for overtaking maneuvers λ_{over}, but only with the simplified observation vector of Eq. (3.3). This does not include the free space variable f, and therefore does not take any information on the required free space into account.

[1]Another reason for approaching may be the missed recognition of the leading vehicle by the driver or a wrong guess of its current velocity. These situations are predicted in the same way.

- The *feasibility of the maneuver* is also represented by a model for overtaking maneuver $\lambda_{over,fs}$, but with the adapted observation vector of Eq. (3.4). Thus, it takes the required free space into account.

A similar prediction as for lane changes is executed using Eq. (4.1) with the two models λ_{over} and $\lambda_{over,fs}$:

$$c_2 = \frac{P(\lambda_{over}|\mathbf{o})}{P(\lambda_{over,fs}|\mathbf{o})} \tag{4.3}$$

High values of c_2 also correspond to high abrupt braking probabilities. It has to be mentioned, that the usage of the two different observation vectors of Eq. (3.3) and Eq. (3.4) requires the training of different models. Thus, different training databases $\mathbf{o}_{train}$ have to be generated for both models.

4.1.3. Evaluation

Evaluation of the described prediction approach depends highly on the specific ADAS feature it is used for. Consider the prediction of lane change maneuvers as an example: For some driver warning features, like a blind spot alert, a prediction at a very early point in time might be appreciated. Usually, this corresponds to a higher rate of incorrectly executed warnings (false positives), where the driver is in fact not going to perform a lane change. For such warning features, those errors may be acceptable. On the other hand, these cases are not tolerable at all for a collision avoidance system performing abrupt braking or steering interventions. Therefore, the general problem is the importance weighting of the true positive rate, the false positive rate and the point in time of the maneuver prediction.

For the evaluation of both use cases, a database with sequences of lane change and abrupt braking maneuvers is built. Maneuvers as depicted in Figure 4.1 and Figure 4.2 are acquired with more than 100 lane change and braking maneuvers each. The used sensor setup for receiving observation data will be described in the following Section 4.2.

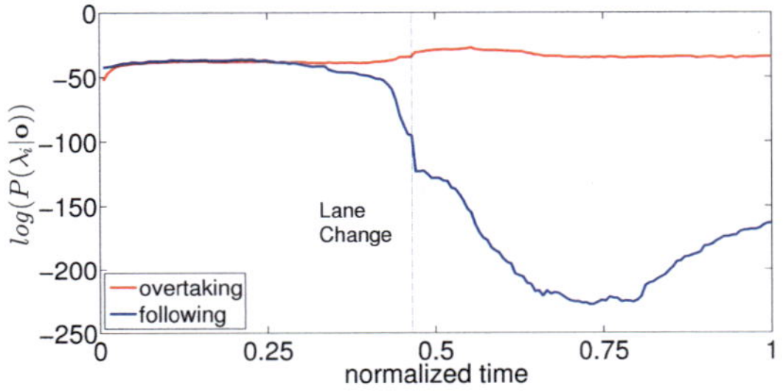

Figure 4.3.: Normalized results of maneuver recognition over all lane change sequences. Normalization is done with respect to point in time of the lane change maneuver; see (Firl et al., 2013)

Evaluation of lane change prediction: The recorded sequences contain different types of lane change maneuvers, thus the stages illustrated in Figure 4.1 are mainly varied in length. Different absolute and relative velocities are also taken into account as well as different relative, longitudinal distances of the vehicles during the execution of the maneuver. The recognition results of the relevant maneuvers are depicted in Figure 4.3, where all sequences are normalized with respect to the points in time of the execution of the lane change (which are labeled manually). Normalization is done using linear time warping (LTW). At the point in time of the maneuver execution a clear divergence of the two maneuver probabilities can be noticed. At the beginning of the sequences, $P(\lambda_{foll}|\mathbf{o})$ is only slightly above $P(\lambda_{over}|\mathbf{o})$, as the first following stage of each sequence varies much in length, or is sometimes almost not present.

The easiest way for evaluating is to set a fixed threshold c_1 and record the points in time of prediction, where values of the ratio $\frac{P(\lambda_{over}|\mathbf{o})}{P(\lambda_{foll}|\mathbf{o})} \geq c_1$ are associated with a maneuver prediction. In Figure 4.4 the results for the complete set of sequences are

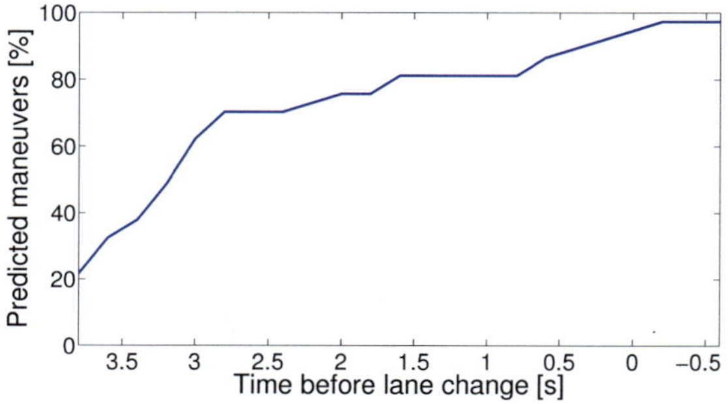

Figure 4.4.: Prediction results for lane change maneuvers with respect to the exact time of the lane change. The used threshold is $c_1 = 1$.

illustrated, using $c_1 = 1$ as threshold. The exact point in time of the lane change maneuver, i.e., when the left tires of the rear vehicle are passing the lane markings, is at $t = 0$. At that time, more than 95% of the maneuvers are predicted. It has to be recognized, that not all maneuvers can be "predicted" at the point in time when the maneuver is already executed due to the probabilistic nature of the prediction approach. A combination with a simpler maneuver recognition (e.g., by observing the lateral offset of the single vehicle in the lane) may compensate this drawback. However, the high potential of this concept is made clear by a prediction rate of about 75% at 2 seconds before the lane change is executed.

As the prediction of lane changes is executed with a specific threshold c_1, the choice of this value is of significant importance. The results with $c_1 = 1$ of Figure 4.4 take only situations into account, where the maneuvers are predicted correctly. That means, that all situations are neglected, where lane changes are wrongly predicted (thus where in fact no lane change maneuver is executed at all). However, these false positives have to be considered to provide a complete evaluation. The ratio of false positive and true positive is shown as a ROC (*Receiver operating characteristic*) curve in Figure 4.5(a). Therefore, a database of $n_{db,lc} \approx 100$ sequences is used, which consists of sequences as illustrated in Figure 4.1 but all contain the first following stage. The following definitions for evaluating the lane change prediction approach are applied:

- *True positives* $\mathrm{TP_{lc}}$: sequences in which a lane change is correctly predicted due to the model probabilities. True positive rate: $\frac{\mathrm{TP_{lc}}}{n_{db,lc}}$
- *False positives* $\mathrm{FP_{lc}}$: sequences in which a lane change is predicted, while both vehicles are still performing a following maneuver and no indication of a lane change is given at all. False positive rate: $\frac{\mathrm{FP_{lc}}}{n_{db,lc}}$

The ROC curve consists of different applied prediction thresholds c_1, which implies both, different $\mathrm{TP_{lc}}$ and $\mathrm{FP_{lc}}$. The desired but usually inaccessible point is the upper left corner at $\mathrm{TP_{lc}} = 1$ and $\mathrm{FP_{lc}} = 0$. Thus, a reasonable trade-off has to be made, depending on the weighting of both values. The threshold $c_1 = 1$ of Figure 4.4 results in $\mathrm{TP_{lc}} \approx 0.85$ and $\mathrm{FP_{lc}} \approx 0.05$, which may be applicable values for some applications. Increasing c_1 will lead to smaller true positive and false positive values, whereas a decreasing of c_1 will result in higher true positive and false positive values. The influence of a higher c_1 value on the prediction time, as shown for $c_1 = 1$ in Figure 4.4, is that the higher the value of c_1 is chosen, the later the prediction of the maneuver is generally be performed.

Evaluation of abrupt braking prediction: For the prediction of abrupt braking maneuvers, the recorded sequences consist of the different stages as illustrated in Figure 4.2. They are mainly varied with respect to the length of the stages, the point in time of the abrupt braking and the driving behavior of the vehicle on the left lane (representing the blockage of that lane). As these sequences are acquired with real vehicles, not all braking

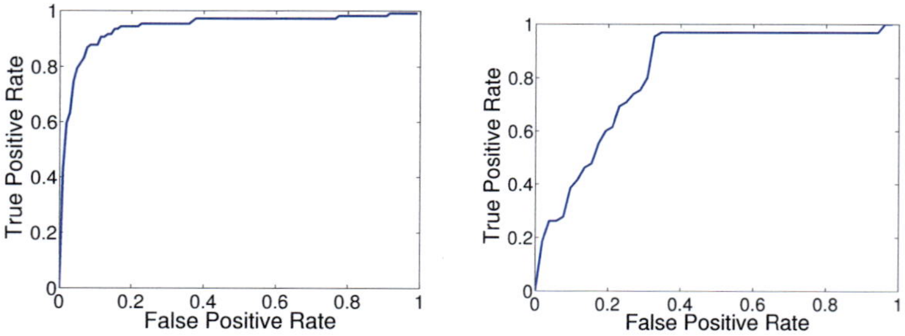

(a) ROC curve for lane change maneuver prediction.

(b) ROC curve for braking maneuver prediction.

Figure 4.5.: ROC curves for lane change and abrupt braking prediction.

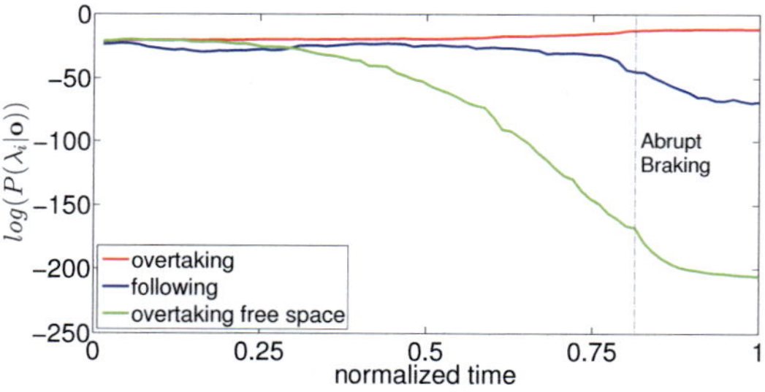

Figure 4.6.: Normalized results of maneuver recognition over all abrupt braking sequences. Normalization is done with respect to point in time of the abrupt braking maneuver; see (Firl et al., 2013)

maneuvers occurring on public streets could be taken into account due to safety reasons. The recognition results for the relevant maneuvers are depicted in Figure 4.6, where the maneuver probabilities are normalized with LTW again. The results for the following maneuver are included for a better interpretation of the graphs. At the point in time of the maneuver execution a clear divergence of the driver's intention (*overtaking*) and the maneuver feasibility (*overtaking free space*) can be noticed.

For evaluating the abrupt braking maneuver prediction, the influence of the vehicle on the left lane on the observation vector (more precise, on the free space variable f) plays a distinctive role. As defined in Section 3.2.4 the used occupancy grid has to be defined for the usage of model $\lambda_{over,fs}$ in Eq. (4.3). To keep the computational complexity rather low, the used grid only consists of one row for both lanes each (compare with Figure 3.10(c)). The length of each row is set to $c_{over} = 10$. As not only the actual position of the vehicle should be taken into account, but also its future course, a prediction method has to be selected. For this evaluation purpose, a constant velocity model is therefore applied, for predicting over a given time horizon of 3 seconds. The more reliable the prediction for the vehicle is to occupy a single cell (i, j), the closer the value $f_{i,j}$ is to zero (reliability refers to be predicted even for a small time interval).

As for the lane change prediction, a straight forward way to analyze the capabilities of the abrupt braking prediction is to use a fixed threshold c_2 and notice the point in time of the maneuver prediction, i.e. where $\frac{P(\lambda_{over}|\mathbf{o})}{P(\lambda_{over,fs}|\mathbf{o})} \geq c_2$ holds. The results when choosing $c_2 = 1$ are illustrated in Figure 4.7. The exact point in time of the braking maneuver is at $t = 0$, i.e. when the vehicle's absolute deceleration is exceeding a predefined threshold. At this point in time almost all maneuvers are predicted but not all of them, as already mentioned in the last paragraph. A more significant value than the one at time $t = 0$, when a "prediction" is in fact no longer of interest, is the prediction rate at 1 second before the braking is executed of $\approx 70\%$.

Similar to Figure 4.5(a), in Figure 4.5(b) the ROC curve for the braking prediction approach is depicted. The used database consists of $n_{db,br} \approx 100$ sequences as illustrated in Figure 4.2, but all contain the first stage, where no vehicle blocks the left lane.

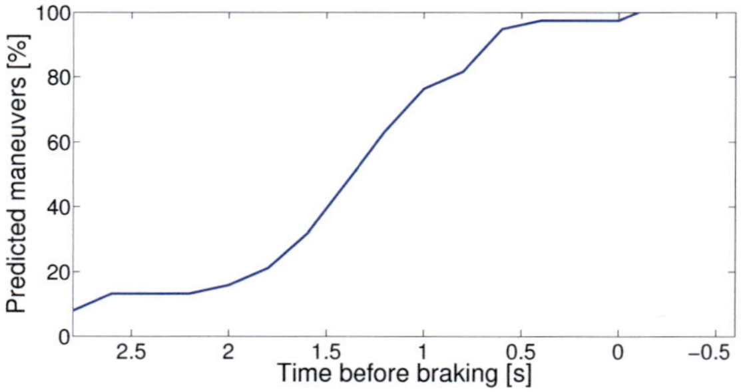

Figure 4.7.: Prediction results for abrupt braking maneuvers with respect to the exact time of the braking. The used threshold is $c_2 = 1$.

This stage is used to consider false positives, as the following definitions hold:

- *True positives* $\mathrm{TP}_{\mathrm{br}}$: sequences in which an abrupt braking is correctly predicted due to model probabilities. True positive rate: $\frac{\mathrm{TP}_{\mathrm{br}}}{n_{db,br}}$
- *False positives* $\mathrm{FP}_{\mathrm{br}}$: sequences in which an abrupt braking is predicted, while both vehicles are still performing a following maneuver and no indication of a braking is given at all. False positive rate: $\frac{\mathrm{FP}_{\mathrm{br}}}{n_{db,br}}$

Thus, the ROC curve consists of different applied prediction thresholds c_2. Depending on the accessed application, different ratios of $\mathrm{TP}_{\mathrm{br}}$ and $\mathrm{FP}_{\mathrm{br}}$ and therefore different values of c_2 may be reasonable. When comparing the curve with the one from the lane prediction approach, a general lower true positive rate with respect to same false positive rates can be noticed. The reasons therefore are manifold, however a dominant one is the modeling of the blocked lane by another vehicle. Thus only very safe braking maneuvers are included in the database, which are usually harder to predict.

4.2. Application for Car-to-X Communications

The approach for predicting lane change and abrupt braking maneuvers has applications in the domain of vehicular safety. There, the assessment and prediction of possible safety critical situations represent requisite information. In the following, the integration of the advocated maneuver recognition and prediction approach into one specific application is presented. After showing the functionality using the sensor setup presented in Section 3.3, in this section the focus is on upcoming Intelligent Transportation Systems (ITS) based on Car-to-X (C2X) communication.

In Section 4.2.1 an introduction to C2X communication and their security aspects is given, focusing on the actual European market. Section 4.2.2 presents the integration of the proposed maneuver prediction approach into the framework for verifying the trustworthiness of mobility data contained in C2X messages.

4.2.1. Motivation: Car-to-X Security

C2X communication, mainly consisting of Car-to-Car (C2C) and Car-to-Infrastructure (C2I) communication, exchange traffic information to increase road safety and traffic efficiency. Therefore the IEEE 802.11p standard of WLAN communication was established in 2010, earmarked by the European Commission (Intelligent Transport Systems, 2010a) and operated in the 5 GHz frequency band. For its usage a lot of standardization work has to be done, especially regarding the used message set and its definition. This is currently performed by the European Telecommunications Standards Institute (ETSI). Based on this message set, first field test trials are conducted, like the European DRIVEC2X (Drive C2X, 2011) and the German sim^{TD} project (sim^{TD}, 2011). The most important message type for this work is the so called Cooperative Awareness Message (CAM), which includes the vehicle's mobility information, mainly consisting of position, speed and heading, and a unique ID of the vehicle[2]. They are sent within intervals from 1 to 100 Hz depending on the current traffic situation and its characteristics (Intelligent Transport Systems, 2010b).

The exchange of CAMs among traffic participants enables a variety of different new ADAS features as well as an optimization of already existing ones. However, security and privacy issues have been identified as a crucial enabler for C2X by the ETSI, as it is typical for open communication methods (Intelligent Transport Systems, 2009). For this purpose, cryptographic methods are usually applied in terms of a Public Key Infrastructure (PKI) (Bißmeyer et al., 2011) using the standard IEEE 1609.2 (Intelligent Transportation Systems Committee, 2006). This enables the usage of digital certificates and signatures, so that the trustworthiness of every CAM message can be authenticated.

Nevertheless, security for C2X using cryptography has been identified to be a necessary but not sufficient instrument to prevent the forging of messages. If an adversary has gained access to secret material of the PKI, he will be able to send messages, which cannot be identified by cryptographic security mechanisms. Therefore, alternative message verification methods become requisite. The presented verification approach to handle that issue (see Section 4.2.2) is motivated by the following description of an attacker on C2X communication networks.

Attacker Model: In the following section an attacker is assumed, who is able to send and to manipulate the information included in the CAMs to the C2X modules. Thus, he has the ability to provoke different safety critical situations, as depicted

[2]Another message type is the Decentralized Notification Message (DENM), which is only sent when special traffic events occur (Intelligent Transport Systems, 2010c).

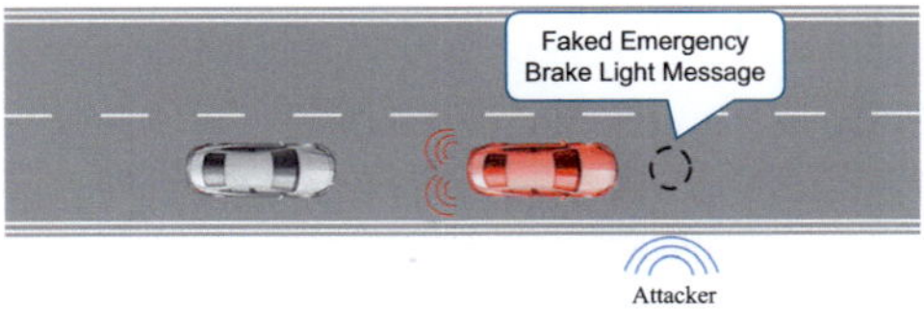

Figure 4.8.: Attacker on C2X communication: Faked Electronic Emergency Brake Light (EEBL) message (Firl et al., 2013).

exemplarily in Figure 4.8. Using a common C2X module, the attacker is able to send authenticated messages, if having access to valid secure key material. By faking a full brake a respective warning message is created assuming a certain risk of collision. If the message is created with manipulated GPS information, any reference position can be stated. Since the attacker is using valid key information, the receiver of those messages will mark them as valid. So cryptography cannot detect such an attacker. In the situation of Figure 4.8 upcoming vehicles will be wrongly alerted by a collision warning, which may result in unnecessary and often dangerous collision mitigation or avoidance maneuvers.

Consequently, additional verification steps have to be performed on receiver side to verify the trustworthiness of C2X messages. In the following, a two-stage verification approach is presented. The basic part consists of a Kalman filter (KF), which is modified by the maneuver prediction component of Section 4.1 of this thesis. It was originally developed during the field trials of sim^{TD}.

4.2.2. Verification Approach for Car-to-X Mobility Data

The basic approach for the verification of incoming C2X CAMs on receiver side (on the *observer car*) consists of two different stages. The first part as presented in (Stübing et al., 2010) mainly consists of a Kalman filter evaluating a whole sequence of messages instead of only single ones. It performs a single step prediction of the vehicle's path (of the *observed car*) and a comparison between this prediction and the received mobility data. An exact definition of the used system states and KF assumptions is given in (Stübing et al., 2011) and is not in the focus of this thesis (see the appendix Section A.3). Afterwards, the second stage performs an adaption of the used filter to increase the reliability of the overall system. Therefore, the maneuver prediction of the last section is applied showing the potential of the proposed modeling concept of this thesis.

Figure 4.9 illustrates the complete verification process of received CAMs and their mobility information. Thus, incoming messages are classified as *Approved*, *Erroneous* or *Neutral*. First, some *Basic Checks* have to be passed based on physical and regulatory boundaries including checks of maximum velocity and message frequency, see (Stübing et al., 2010) for a complete list. Only if all checks are passed (D1), the first verification stage is queried, otherwise the message is marked as *Erroneous*.

For new vehicles, which do not have a known vehicle ID, a new tracker is instantiated on the host car including a KF. Therefore, a margin check is carried out assuming new vehicles to appear somewhere at the outer border of the host vehicle's communication range. If a vehicle's CAM is received for the first time, no reliably prediction can be performed. Thus, in the following it is assumed that the vehicle has a known ID, so (D2) is passed with *yes* in Figure 4.9. If the vehicle ID of the received message is already known, the Kalman prediction phase is executed using a constant velocity motion model. The difference Δy_t between received mobility data and Kalman predicted state at the current time t is determined using a transition matrix.

This difference serves as a measure for the plausibility of the mobility information of the received message. The smaller the

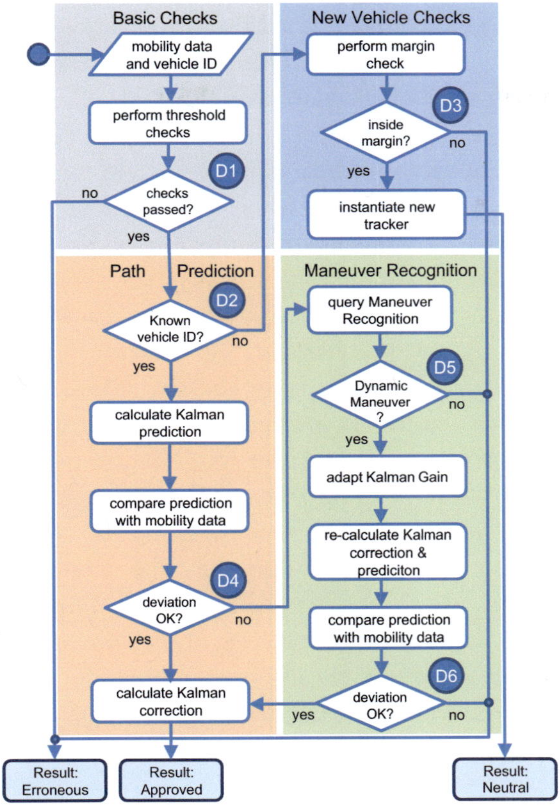

Figure 4.9.: Verification flow for C2X mobility data using single step path prediction and probabilistic maneuver prediction (Firl et al., 2013).

difference is, the better the received data matches the model assumption of the KF. This is part of the steps before (D4) at the left side of Figure 4.9. Therefore, a predefined acceptance threshold is used based on expected GPS errors of $1-3$ meters. Thus, only the norm of the 2-D position information is used for the evaluation of the deviation in (D4). If it is passed, the Kalman correction phase is executed and the message is classified as *Approved*. Assuming an attacker as depicted in

Figure 4.8 who is not using a valid motion model for sending a sequence of faked messages will result in not approved messages when applying the verification step (D4).

For most traffic situations the used acceptance threshold results in an acceptable performance, i.e. most messages of correctly sending vehicles are marked as *Approved*[3]. However, in situations where the motion assumption of the KF does not match the driving behavior accurately, the deviation exceeds the threshold more often. These situations are identified to be characterized by suddenly occurring dynamic changes in the motion of the vehicle. Especially the two situations named in the last section, suddenly performed lane change maneuvers and abrupt brakings, exceed the modeling capabilities of the used KF. Therefore, an adaption of the KF has to be applied in these cases as described in the following.

Assuming an exceeding of the AT, the maneuver recognition stage is triggered in (D4) of Figure 4.9. Since an adaption of the KF should only be done in situations where the motion model does not match the current driving behavior, the maneuver prediction approach is executed. Therefore, the relevant maneuver probabilities are calculated first, i.e. $P(\lambda_{over}|\mathbf{o})$ and $P(\lambda_{foll}|\mathbf{o})$ for a lane change maneuver and $P(\lambda_{over,fs}|\mathbf{o})$ and $P(\lambda_{over}|\mathbf{o})$ for an abrupt braking maneuver, respectively. If one of these maneuvers is predicted successfully using criteria Eq. (4.2) and Eq. (4.3), an adaption of the used Kalman Gain is conducted. The used observation vector $\mathbf{o}$ from Eq. (3.4) is acquired by calculating the relative kinematics from the received mobility data of the CAM and the information of the observer vehicle. The KG weights the predicted system state with the actual measurement. As it is assumed that the motion model fails to explain the vehicle's driving behavior accurately,

[3] An evaluation of the verification approach used in common traffic scenarios can be found in (Firl et al., 2013), where the real driving behavior matches quite well with the motion assumption of the KF.

the actual system state is corrected towards the measurement, depending on the maneuver prediction result. The more precise the prediction of the maneuver is (i.e., the greater the ratio of Eq. (4.2) or Eq. (4.3) is), the larger the correction towards the measurement will be.

By predicting a dynamic maneuver and adapting the Kalman Gain the received mobility data is not automatically marked as *Approved*. However, the adapted Kalman Gain is applied by reversing the previous prediction and correction phase and a recalculation using the new Kalman model. The enhanced corrected state leads to a new prediction, closer to the received mobility data. Thus the old deviation Δy_t is decreased resulting in $\Delta y_{t,new}$. The same acceptance threshold is applied in (D6) of Figure 4.9 again and the message will be classified as *Approved* if it succeeds. Otherwise it will be marked as *Erroneous*, finally.

The main advantage of this adapted verification approach is that fewer messages are falsely marked as *Erroneous* without increasing the acceptance threshold AT. This would have undesired security consequences, since a higher acceptance threshold simplifies the sending of verified messages from an attacker. In the following section exemplary results are shown to demonstrate the benefit of this new message verification approach.

4.2.3. Results of Message Verification

The mobility data verification framework is fully implemented as a Java/-OSGI bundle into the C2X communication system used for the sim^{TD} field trials. See also A.2 for more information on this, including implementation details. For evaluation purposes, three vehicles are used, which are equipped with communication components of sim^{TD}. Two vehicles are used for performing the specific traffic maneuver and the third vehicle is used for representing the free space on the left lane as de-

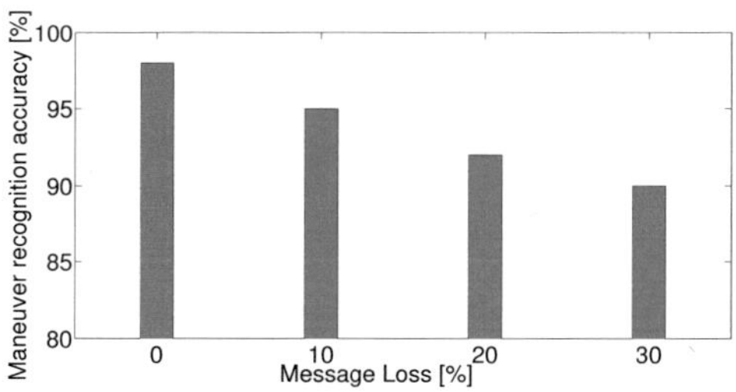

Figure 4.10.: Maneuver recognition for different message loss rates (Firl et al., 2013).

picted in Figure 4.2. The same database as already mentioned in Section 4.1.3 is used consisting of more than 100 sequences of dynamic lane change and abrupt braking maneuvers each.

The first assessment is focusing on the general performance of the proposed approach handling different message loss rates. As C2X communication is assumed to be lossy, this sensitivity is of particular interest. In general the average deviation Δy_k increases with an increased message loss rate as well as the number of peaks, i.e. the number of messages exceeding the acceptance threshold AT. These messages result in false positives since all messages represent correct vehicle's information. In Figure 4.10 the proposed maneuver prediction approach is evaluated with respect to message loss rates up to 30%. For each sequence of the database and for each message loss rate about 1000 variants of message losses are generated randomly. Even for high loss rates of 30% the accuracy of the maneuver prediction component remains above 90%. This corresponds to a low false negative rate, i.e., only few peaks occur during a dynamic maneuver while not correctly predicting the corresponding maneuver by the probabilistic recognition component.

In the following, the benefit of the proposed two-stage verification approach is presented exemplary for one sequence of a dynamic lane change and an abrupt braking maneuver each.

In Figure 4.11 the results of the proposed message verification approach for one sequence including a dynamic lane change maneuver are illustrated. The graph at the top shows the isolated results of the KF (without the adaption of the maneuver prediction) and the deviation Δy_t. The AT value is set to 1 meter. While the x-axis denotes the number of CAM message received by the host vehicle, the y-axis shows the behavior of Δy_t. As already assumed, the deviation becomes maximal during the third stage of the maneuver, where the lane change is performed. There, peaks up the 3 meters can be observed, exceeding the security threshold multiple times (messages $37 - 52$). These messages don't pass the check (D4) of Figure 4.9 and would be discarded if not using further verification stages or increasing the acceptance threshold significantly to approximately 3 meters.

In the middle part of the image, the results of the maneuver recognition are depicted, which are crucial for predicting lane changes. Both curves reflect very well the current status of the vehicle during the complete maneuver in terms of probabilities for an overtaking and a following maneuver, respectively. The point in time when the maneuver recognition component is triggered first is at message 37 where $\Delta y_t > AT$ holds for the first time. Since the probability $P(\lambda_{foll}|\mathbf{o})$ is decreased already, a lane change maneuver can predicted correctly according to Eq. (4.2).

At the bottom of Figure 4.11 the values of $\Delta y_{t,new}$ are shown. Since a lane change maneuver is correctly assumed at all points in time when $\Delta y_t > AT$ holds, the Kalman Gain is adapted such that the first peak at message 37 is successfully decreased below 1 meter. Thus, the message is correctly classified as *Approved.* The subsequent peaks are handled similarly, such that no messages are wrongly marked as *Erroneous* at all. Keeping in mind that this would only be possible by increasing the security threshold AT resulting in a decreased overall security level outlines the benefit of the adapted message verification framework of Figure 4.9.

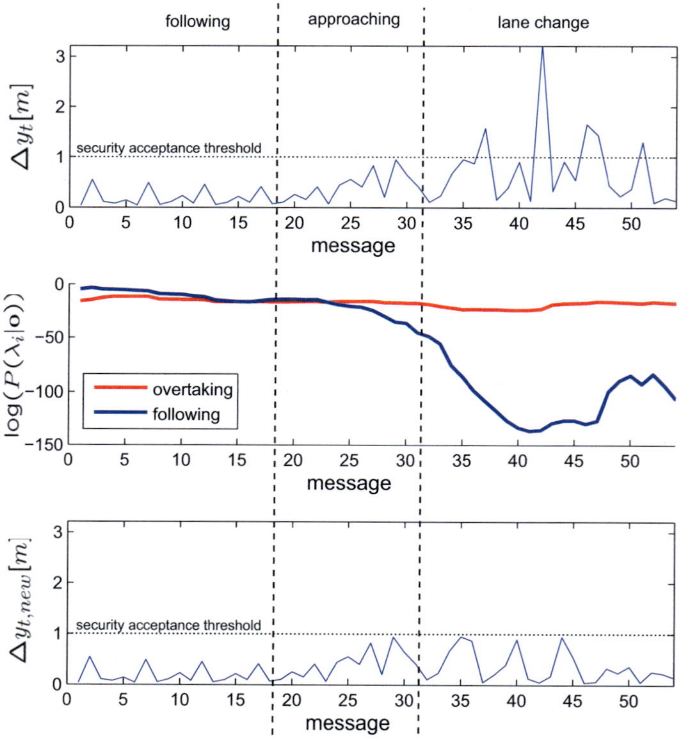

Figure 4.11.: Lane change maneuver - Top: Kalman deviation Δy_t - Middle: log-likelihood of maneuver recognition - Bottom: Kalman deviation $\Delta y_{t,new}$ with adapted Kalman gain

Similar to the evaluation of the lane change prediction, the results of the verification approach for one sequence including an abrupt braking are presented in Figure 4.12. The different stages of the sequence are similar to the lane change sequence, but this time the lane change cannot be executed due to insufficient free space on the left lane caused by the third involved vehicle. That is why the vehicle has to perform a hard braking maneuver to avoid a collision. This leads to high deviations Δy_t in the last stage of the sequence. The resulting peaks can be seen in the curve at top of the figure, appearing at message 70 for the first time.

In the middle part of Figure 4.12 the corresponding maneuver recognition results are shown. They are queried when the deviation Δy_t exceeds the AT value. At this point in time, the driver's intention to overtake the vehicle ahead (overtaking, no free space consideration: λ_{over}, red curve in Figure 4.12) clearly diverges from the real feasibility of the maneuver (overtaking, free space consideration: $\lambda_{over,fs}$, green curve in Figure 4.12). Thus, the abrupt braking prediction is able to be performed successfully according to Eq. (4.3).

Therefore, the adaption of the Kalman Gain is done accurately resulting in a higher weighting of the received message with respect to the Kalman prediction. The new deviation $\Delta y_{t,new}$ is depicted by the graph at the bottom of the figure. Using the adapted KF results in a deviation, which does not exceed the same acceptance threshold value anymore.

The evaluation of both sequences, for the lane change manoeuver and for the abrput braking, shows an decreased number of false positives when applying the adapted message verification approach. However, an evaluation of true positives is still missing, i.e., situations where an attacker is sending faked messages. On the one hand, this is not the scope of this thesis and was evaluated in previous works of the sim^{TD} project. On the other hand, this evaluation cannot be performed by persons, who are aware of the message verification approach presented in this chapter. Attackers who has access to the secret material of the PKI and are aware of the applied message verification concept will not be identified by this approach.

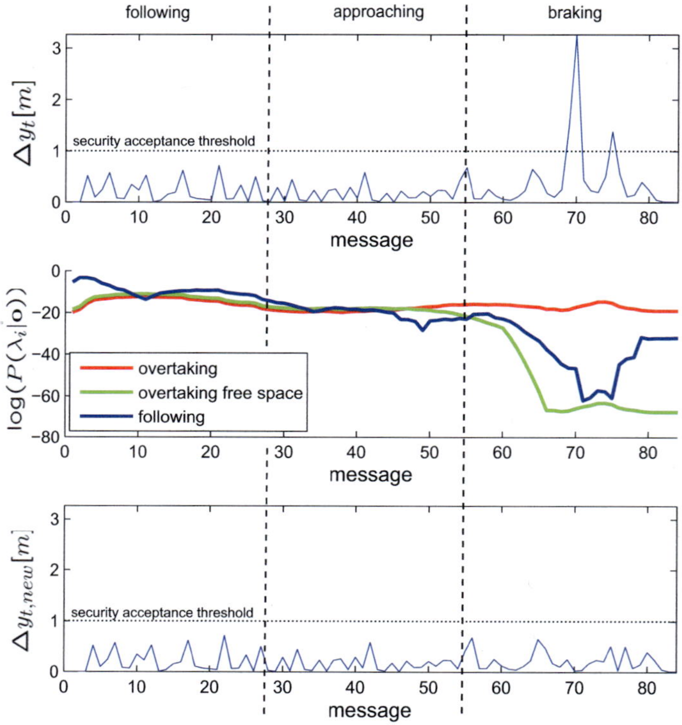

Figure 4.12.: Braking maneuver - Top: Kalman deviation Δy_k - Middle: log-likelihood of maneuver recognition - Bottom: Kalman deviation $\Delta y_{k,new}$ with adapted Kalman gain

4.3. Summary

In this chapter evaluation concepts for the maneuver recognition framework of Chapter 3 were outlined. As typical for most works focusing on situation assessment, the main difficulty is the dependency on the accessed feature. Therefore, a general evaluation method was presented based on the prediction of single dynamic maneuvers. Furthermore, the proposed concept was integrated into an existing software architecture to show the benefit for one real application exemplary. Furthermore, by using C2C communication messages as observation data, this chapter pointed out the independence of the HMM based framework of this thesis on the used sensor setup (see also Figure 1.1).

In Section 4.1 the performance of the probabilistic framework was tested by predicting dynamic maneuvers in terms of lane changes and abrupt brakings using the maneuver probabilities $P(\lambda_i|\mathbf{o})$. This is done by observing the ratio between two models for each maneuver. Two different evaluations were done using a database of about 100 sequences for each maneuver. The first one evaluates the prediction time with respect to the point in time of the real maneuver execution. Since different ADAS features require a great robustness concerning false positives, these were treated by ROC curves. The results show the high potential benefit of the proposed prediction approach for different features.

The integration into a real vehicle application was presented in Section 4.2 focusing on the security aspects of Car-to-X communication networks. An already existing approach for verifying the trustworthiness of incoming C2C communication messages was extended by the proposed maneuver prediction approach. Since the original framework lacks the handling of messages sent during dynamic traffic maneuvers, the proposed concept is able to solve these shortcomings. Therefore, the Kalman Filter used for predicting incoming mobility data was adapted appropriately resulting in an increased overall security level.

5. Maneuver adaptive Trajectory Prediction

In Chapter 3 and Chapter 4 a concept for recognizing traffic maneuvers in a probabilistic manner was presented. Furthermore, an integration into an entire ADAS architecture was outlined by means of C2X message verification. However, for many ADASs the most crucial information for executing any warning or intervention strategy are the trajectories of relevant traffic participants in the current and future traffic scene. The more accurate the predictions of these trajectories are, the more robust the system may perform. The proposed maneuver prediction system of Chapter 4 does not provide such information.

As mentioned in Section 4.1, the maneuver modeling using HMMs does not support plausible predictions of single traffic participants since no kinematic motion model is integrated. The predicted states received by sampling the observation model would result in non-conformable trajectories as the different entities of the state vector (e.g., position and speed) are not related directly to each other in the model. Therefore, a concept for predicting vehicle trajectories is presented in this chapter, taking the occurrence and probabilities of traffic maneuvers into account. Aspects of the following are the results of the constructive and cooperative work of (Omerbegovic, 2012) and were published in (Omerbegovic and Firl, 2013).

In Section 5.1 a motivation of the maneuver adapted trajectory prediction is given, pointing out some drawbacks of commonly used state of the art predictors. Afterwards, the proposed approach is described in detail in Section 5.2 including some basic information on the used theoretical concepts.

Results with simulated as well as with real-world data are finally presented in Section 5.3.

5.1. Motivation

Most prediction methods for the trajectories of traffic participants are usually taking into account:

- the information on the **current vehicle's state** (e.g., position, velocity),
- information on **past vehicle's state**, gathered and tracked over a certain period of time,
- a kinematic **motion model**, using some simplifying assumptions on vehicle's motion (e.g., constant acceleration, constant yaw rate).

The most popular predictor therefore is the Kalman Filter (KF) using different motion assumptions and different extensions of the filter itself (e.g., Extended KF and Unscented KF).

Furthermore, information on the **road geometry** with different lanes is often included to adapt the predicted trajectory on the current static traffic scenario. Adaptions on **dynamic traffic participants** are usually added by assuming collision-free trajectories. However, while achieving reliable prediction results in many common traffic situations, the model capabilities reach their limits in dynamic scenarios, involving multiple traffic participants. Thus, accurate predictions are only possible over a short time horizon.

The contribution of this chapter is therefore to provide a trajectory prediction approach with the following properties:

- handling traffic situations involving multiple, interacting traffic participants on extra urban roads,

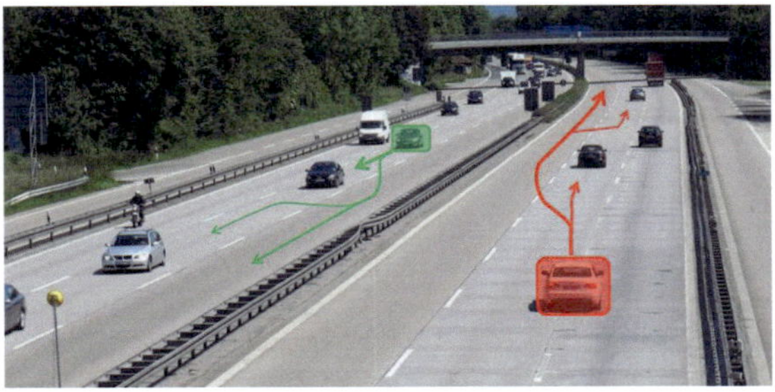

Figure 5.1.: Example of possible results of trajectory prediction of two vehicles on a German Autobahn considering maneuvers of multiple traffic participants. Different line widths indicate different probabilities of the trajectories.

- providing trajectory predictions over a large time horizon of multiple seconds,
- providing probabilistic predictions including multiple possible trajectories with associated probabilities.

In Figure 5.1 an example of possible prediction results on an extra urban road is illustrated.

5.2. Trajectory Prediction Approach

The basic idea of predicting trajectories in traffic situations where traffic maneuvers of multiple vehicles are performed is depicted in Figure 5.2. In situations, where the relevant vehicle is driving without any interaction to other traffic participants, conventional prediction methods may be applied. If the start of an interaction is observed, the prediction system switches into another mode, considering different maneuvers and their probabilities. In the following, this second mode will be discussed in detail, since single object trajectory prediction is not the scope of this thesis. The transition from one mode into the other may be realized in different ways, depending highly on the used sensor setup. The situation adaptive stage may be exemplarily triggered, if:

- at least one vehicle is in the sensor's detection range,
- at least one vehicle is within a predefined distance to the relevant vehicle,
- at least one maneuver probability of an interaction with any other traffic participant exceeds a predefined threshold.

Since the following approach is independent of the used sensor setup for retrieving input data, the transitions between both stages is not analyzed. Hence, it is assumed that the second stage was already triggered and maneuver probabilities to at least one other vehicle are already computed.

Before going into detail of the proposed approach, some basic definitions and their context have to be pointed out. A discrete trajectory is given by a set of samples and corresponding time stamps. As the time base is assumed to be equidistant, a trajectory can be written as:

$$\mathbf{x}_{1:N} = \{x_t\}_{t=1}^{N}, \tag{5.1}$$

where x can be either scalar or multidimensional. For vehicle trajectory purposes, the state vector x_t contains at least the position of the vehicle in some coordinate system at time t. As trajectories depending on traffic maneuvers are discussed in the following, the same state vector as defined as the observation vector o_t in Chapter 3 is used, see Eq. (3.3). Thus, the vehicle and lane oriented coordinate system is used.

The task of situation- (or maneuver-) dependent trajectory prediction is formulated as follows: Given (i.e., observed) is a trajectory $\mathbf{x}_{1:N}$, consisting of N data samples in a vehicle and

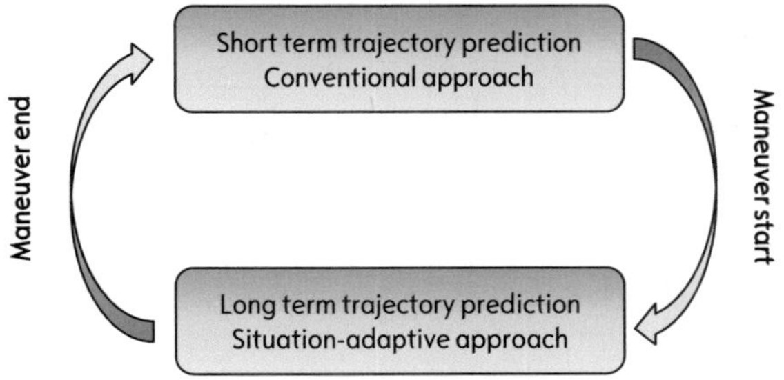

Figure 5.2.: Close loop system for vehicle's trajectory prediction.

lane oriented coordinate system referring to the interaction of a traffic participant pair[1]. Furthermore, the probabilities of the maneuver recognition for the observation sequence $\mathbf{x}_{1:N}$ for all models λ_i (in this thesis: overtaking, following and flanking) are computed, see Section 3.3: $P(\lambda_i|\mathbf{x}_{1:N})$ $(i = 1,\ldots,3)$. The results of the prediction approach presented in the following have to be twofold:

- A set of trajectory predictions have to be determined:

$$\hat{\mathbf{x}}^{(i,k)}_{N+1:N+\Delta N}, \quad i = 1,2,3, \quad k = 1,\ldots,K^i \tag{5.2}$$

  ΔN denotes the prediction horizon, K^i the number of trajectories predicted for the i-th maneuver and (i,k) the index of the k-th prediction of the i-th maneuver.

- Furthermore, the probability for each predicted trajectory given the observed trajectory $\mathbf{x}_{1:N}$ has to be computed:

$$P(\hat{\mathbf{x}}^{(i,k)}_{N+1:N+\Delta N}|\mathbf{x}_{1:N}) \tag{5.3}$$

[1] In general, there may be more than one other traffic participant, a vehicle is interacting with. However, it is assumed in this thesis, that the relevant vehicle is only interacting with one other participant, what reflects most common traffic scenarios on extra urban roads. Otherwise, resulting trajectories from more than one vehicle pair have to be merged afterwards.

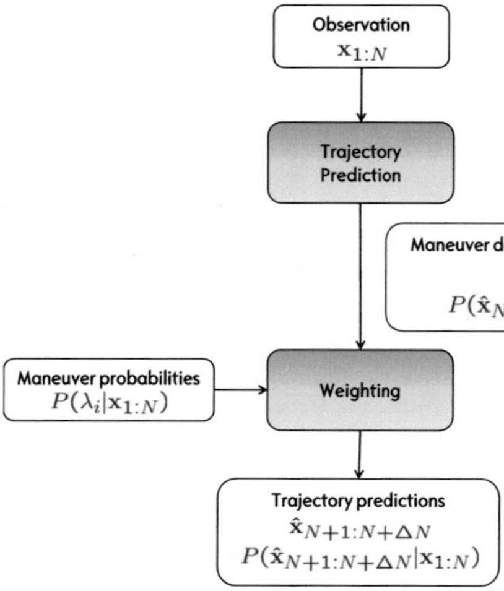

Figure 5.3.: Concept for maneuver adaptive vehicle's trajectory prediction.

The concept for predicting trajectories using the probabilities of recognized maneuvers is depicted in Figure 5.3. For the input observation sequence $\mathbf{x}_{1:N}$ a prediction stage is triggered for every maneuver λ_i, which are assumed to be successively given. Thus, for every maneuver, a set of predictions $\hat{\mathbf{x}}_{N+1:N+\Delta N}$ are computed, including their probabilities $P(\hat{\mathbf{x}}^{(i,k)}_{N+1:N+\Delta N}|\mathbf{x}_{1:N},\lambda_i)$, for a given observation sequence and maneuver. How to achieve this will be outlined in this section later on.

To take the computed maneuver probabilities $P(\lambda_i|\mathbf{x}_{1:N})$ into account, a second step is required to weight all trajectories dependent on the assumed maneuver. This results in the probabilities $P(\hat{\mathbf{x}}^{(i,k)}_{N+1:N+\Delta N}|\mathbf{x}_{1:N})$ for every prediction by a simple multiplication. Mathematical more precise, the resulting probabilities are $P(\hat{\mathbf{x}}^{(i,k)}_{N+1:N+\Delta N},\lambda_i|\mathbf{x}_{1:N})$, applying Bayes' theorem.

Thus, the most challenging part is to predict a set of trajectories for a given input sequence $\mathbf{x}_{1:N}$ assuming a specific maneuver λ_i. In this thesis, a method based on the comparison of different reference trajectories and a higher-level concept of *Case-based reasoning* is presented.

5.2.1. Case-Based Reasoning

Case-based reasoning (CBR) is a method to solve and to reason about problems, when a general analysis of the underlying system is lacking, see (Kolodner, 1992) and (Perner, 2007). Therefore, the actual problem is interpreted as a *case*, which has to be solved by using a database containing previously solved cases including their *solutions*. Generally spoken, CBR consists of the steps *Retrieve*, *Reuse*, *Revise* and *Retain*. Besides different application areas as understanding and solving customer service problems (help desk), also some classification works for automobile applications are published, as in (Vacek et al., 2007). In this work CBR is used to classify different traffic situations for cognitive automobiles.

Adapting the principle of CBR to the prediction of vehicle's trajectory in traffic maneuvers is depicted in Figure 5.4. The interpretation of the case to be solved is the current observation sequence $\mathbf{x}_{1:N}$, which has to be predicted into future. The required predictions are the missing solution to this case. The prediction for this sequence follows the general CBR cycle as follows:

1. **Retrieve:** The first task is to retrieve similar cases, i.e. trajectories $\mathbf{y}^{(j)}_{1:N_j}$, of the repository, which has to be built up for every maneuver (see Section 5.2.3). The most crucial challenge therefore is to define an adequate similarity measure, what is discussed in Section 5.2.2. To use these cases for prediction purposes, $N_j > N$ has to be required,

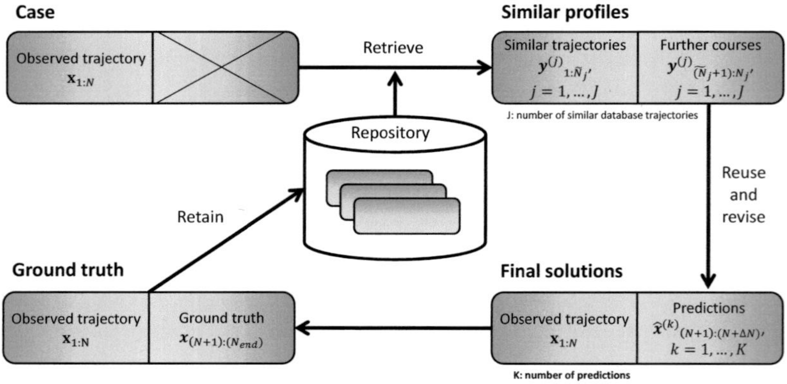

Figure 5.4.: Case-based reasoning for vehicle's trajectory prediction.

splitting the trajectory $\mathbf{y}^{(j)}$ into two parts: the first one indicates the similar part to $\mathbf{x}_{1:N}$ (indicated as $\mathbf{y}^{(j)}_{1:\tilde{N}_j}$) and its further course $\mathbf{y}^{(j)}_{(\tilde{N}_j+1):N_j}$. The number of selected similar trajectories is indicated as J.

2. **Reuse & Revise:** In this stage the selected trajectories $\mathbf{y}^{(j)}_{(\tilde{N}_j+1):N_j}$ are used to draw conclusions on the required predictions $\hat{\mathbf{x}}^{(k)}$. This is presented in detail in Section 5.2.3 with all needed intermediate steps.

3. **Retain:** This last stage is executed, when the complete maneuver was performed, so that the ground truth of the complete sequence $\mathbf{x}$ was already performed. At this time, the complete sequence may be retained to the repository. Since this step is not the focus of this thesis, it is only remarked, that this enables the prediction system to be adapted over time. As an example, it is able to adapt the system to specific drivers, if a large number of trajectories of them are stored and may be reused in future for better prediction results. However, retaining trajectories in the

repository requires adequate strategies to prevent, that too many and too similar trajectories are stored in it, or to prevent the storage to grow exceedingly.

5.2.2. Similarity Measure for Trajectories

In the following the retrieve stage of the proposed CBR cycle is discussed in detail, which depends especially on the definition of a similarity measure to compare the given observation sequence $\mathbf{x}$ with the repository. Basic requirements therefore are:

- handling of different spatio-temporal starting points of the trajectories,
- handling of different lengths of the trajectories,
- handling of different sampling rates,
- robustness to sensor noise and outliers,
- robustness to translations and rotations.

Different similarity measures for comparing trajectories may be applied. In (Hahn et al., 2008) the *Levenshtein Distance on Trajectories* ($LDT(\mathbf{x},\mathbf{y})$) is used, extending the *Levenshtein distance* LD to d-dimensionality. Thus, the number of required operations (insert, delete, replace) are counted to transform $\mathbf{x}$ into $\mathbf{y}$.

The Dynamic Time Warping (DTW) distance, extensively used in speech recognition tasks in the past, is applied to extended application areas of the comparison of time series (Berndt and Clifford, 1994). In (Vlachos et al., 2004) an extension of DTW is presented, archiving translation, scale and rotation invariance. A comparison of different similarity measures is presented in (Zhang et al., 2006), including the DTW distance and some Euclidean distance based measures.

However, most of these approaches seem to be lacking in coping with sensor noise, see (Vlachos et al., 2005). A possible approach to handle this drawback may be the *Longest Common Subsequence* (LCS) model. Originated in the field of string matching, the LCS computes the longest common substring of two input string sequences. In (Hermes et al., 2009), an adaption to the task of comparing vehicle trajectories is presented, which motivates the usage of LCS in this thesis.

The LCS is defined for matching two trajectories $\mathbf{x}$ and $\mathbf{y}$ of lengths N_x and N_y by the recursive formulation:

$$lcs(\mathbf{x}_{1:N_x}, \mathbf{y}_{1:N_y}) = \begin{cases} 0, & \text{if } N_x = 0 \vee N_y = 0 \\ lcs(\mathbf{x}_{1:(N_x-1)}, \mathbf{y}_{1:(N_y-1)}) + s(\mathbf{x}_{N_x}, \mathbf{y}_{N_y}), & \text{if } s(\mathbf{x}_{N_x}, \mathbf{y}_{N_y}) \neq 0 \\ \max\{lcs(\mathbf{x}_{1:N_x}, \mathbf{y}_{1:(N_y-1)}), lcs(\mathbf{x}_{1:(N_x-1)}, \mathbf{y}_{1:N_y})\}, & \text{otherwise} \end{cases} \tag{5.4}$$

The function $s(x,y)$ is a similarity measure of two sequence points x and y. For string matching purposes, where the length of the longest substring has to be found, $s(x,y) = \delta_{x,y}$ (Kronecker delta). For the comparison of trajectories, this is usually not sufficient, since the similarity of the two trajectory points have to be taken into account, too. Thus, a decision criterion has to be defined, either x and y should be matched to each other or not. In this thesis the definition of s introduced in (Hermes et al., 2009) is used. Let D be the dimension of x and y (which corresponds to the dimension of the used observation vector):

$$s(x,y) = \begin{cases} 0, & \text{if } \exists d \in [0,D] : |x^{(d)} - y^{(d)}| > \epsilon^{(d)} \\ \frac{1}{D} \sum_{d=1}^{D} \left(1 - \frac{|x^{(d)} - y^{(d)}|}{\epsilon^{(d)}}\right), & \text{otherwise} \end{cases} \tag{5.5}$$

with the crucial decision threshold $\epsilon \in \mathbb{R}^D$. The choice of this parameter is a non-trivial task, since it indicates if two points of the trajectories are interpreted as *similar* or not. One solution as presented in (Vlachos et al., 2005) is the usage of the minimal standard deviation std of both trajectories, i.e.

$$\epsilon^{(d)} = \min\left(\operatorname{std}(\mathbf{x}^{(d)}), \operatorname{std}(\mathbf{y}^{(d)})\right) \tag{5.6}$$

to take the general course of the trajectories into account. Since it turned out that this results in non-plausible effects, in this work ϵ is set manually for each dimension using numerous testing sequences. See (Omerbegovic, 2012) for more detailed information on that.

To transform the LCS value $\operatorname{lcs}(\mathbf{x}, \mathbf{y})$ into a distance measure the following mapping is applied:

$$d_{\text{lcs}}(\mathbf{x}_{1:N_x}, \mathbf{y}_{1:N_y}) = 1 - \frac{\operatorname{lcs}(\mathbf{x}_{1:N_x}, \mathbf{y}_{1:N_y})}{\max(N_x, N_y)}, \tag{5.7}$$

holding the distance properties:

- $d_{\text{lcs}} \geq 0$,
- $d_{\text{lcs}}(\mathbf{x}_{1:N_x}, \mathbf{y}_{1:N_y}) = 0 \iff \mathbf{x} = \mathbf{y}$,
- $d_{\text{lcs}}(\mathbf{x}_{1:N_x}, \mathbf{y}_{1:N_y}) = d_{\text{lcs}}(\mathbf{y}_{1:N_y}, \mathbf{x}_{1:N_x})$.

The normalization in Eq. (5.7) using the maximum length of both trajectories ensures the definite property. One of the main advantages of this definition is that it ensures the interpretation as a probability measure. This probabilistic interpretation is one of the requirements of the complete trajectory prediction approach of this chapter. Implementation of d_{lcs} is usually done by dynamic programming, which ensures an effective reuse of intermediate results during the recursion.

In the following Section 5.2.3, the complete procedure of predicting trajectories using the proposed CBR-based approach is presented using one exemplary situation.

5.2.3. CBR Cycle for Trajectory Prediction

Before executing the different stages of the prediction approach depicted in Figure 5.4, an appropriate database has to be built up. For this purpose, the same sequences as for training the HMMs of Chapter 3 are used, consisting of simulated trajectories using the simulation tool CarMaker®(see Figure 3.5). The resulted database consists of trajectories $\mathbf{y}$ of different length. The state vector of the database trajectories only consists of the relative longitudinal and lateral position between the two vehicles, and the relative velocity (referring to the observation vector Eq. (3.3), the relative acceleration is neglected).

In the following the different stages of the CBR cycle are described at one exemplary, simulated trajectory indicating an overtaking maneuver. Since overtaking situations imply most difficulties for prediction purposes, the focus is set on this maneuver. However, the same procedure is done for following and flanking maneuvers, which is presented in the following Section 5.3. The used situation represents the beginning of an overtaking maneuver with two lane changes, where the first lane change is currently starting at the point in time when the prediction is triggered (see Figure 5.5).

The first step is to **retrieve** similar trajectories from the database. For this purpose, the LCS distance $d_{\text{lcs}}(\mathbf{x}_{1:N_x}, \mathbf{y})$ of the input trajectory $\mathbf{x}_{1:N_x}$ to all database sequences is computed. Similar trajectories are selected, due to a predefined threshold c_{sim}, i.e., a trajectory $\mathbf{y}^{(j)}$ of length N_j of the database is selected, if

$$d_{\text{lcs}}(\mathbf{x}_{1:N_x}, \mathbf{y}^{(j)}_{1:N_j}) < c_{sim} \tag{5.8}$$

holds. Besides the selection of the threshold c_{sim}, the values of ϵ (i.e., $\epsilon(d_x)$, $\epsilon(d_y)$, $\epsilon(v_{rel})$) of the LCS calculation for each dimension are of utmost importance (see Eq. (5.5)). Both variables determine the number J of selected trajectories $\mathbf{y}^{(j)}_{1:N_j}$. In Figure 5.5 examples of different values of ϵ are shown, where the decision criterion is set to $c_{sim} = 1$. In Figure 5.5(b), where all values of ϵ are smaller than in Figure 5.5(a), it is clearly observable, that not all trajectories $\mathbf{y}^{(j)}$ were selected, which paths are most similar to the ground truth $\mathbf{x}_{(n+1):N_{end}}$ (comparing to the complete trajectory database). Since this is not a reasonable characteristic of this approach, the threshold values of ϵ should not be set too small. An overall best solution for the choice of the parameters is not possible, since it depends highly on the requirements on calculation time and memory. The larger the value of ϵ in one dimension is, the more trajectories are selected of the database due to variations in this value. For this thesis numerous simulations with different sets of threshold values ϵ and c_{sim} are performed to find an adequate parameter selection. In the following, the parameters of Figure 5.5(c) are used, as it results in a reasonable number of trajectories with respect to the used databases.

After associating the trajectories $\mathbf{y}^{(j)}_{1:N_y}$ $(j = 1, \ldots, J)$ of the database to the given input sequence $\mathbf{x}_{1:N}$, a probabilistic weight has to be computed for all of them:

$$P(\mathbf{y}^{(j)}_{1:N_y} | \mathbf{x}_{1:N}, \lambda_i) := \frac{1 - d_{\text{lcs}}(\mathbf{y}^{(j)}_{1:N_y}, \mathbf{x}_{1:N})}{\sum_{l=1}^{J} 1 - d_{\text{lcs}}(\mathbf{y}^{(l)}_{1:N_y}, \mathbf{x}_{1:N})}, \tag{5.9}$$

where $\lambda_i = \lambda_{over}$ indicates the given overtaking maneuver for the current situation.

The last step of the retrieve stage is, that the selected trajectories $\mathbf{y}^{(j)}_{1:N_y}$ have to be split into two parts. The first one indicates the similarity to the input sequence $\mathbf{x}_{1:N}$. This is in-

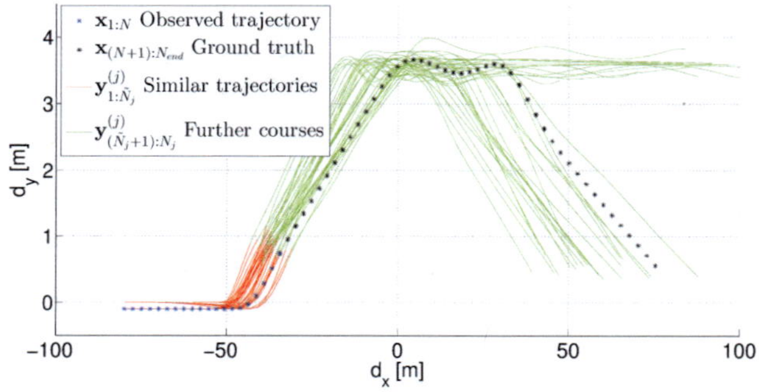

(a) $\epsilon = (10 \text{ m}, 0.5 \text{ m}, 15 \text{ km/h})$

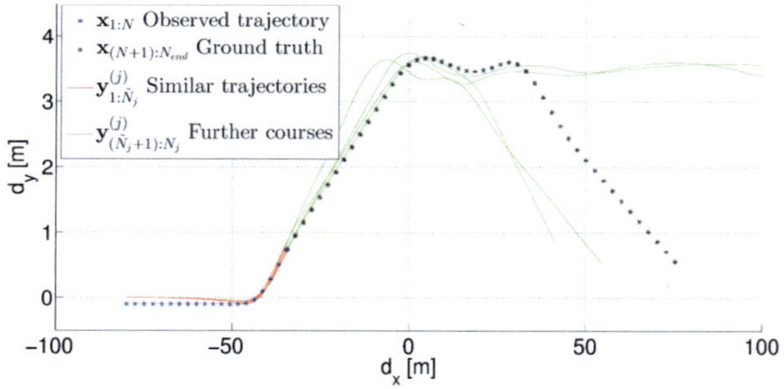

(b) $\epsilon = (3 \text{ m}, 0.1 \text{ m}, 5 \text{ km/h})$

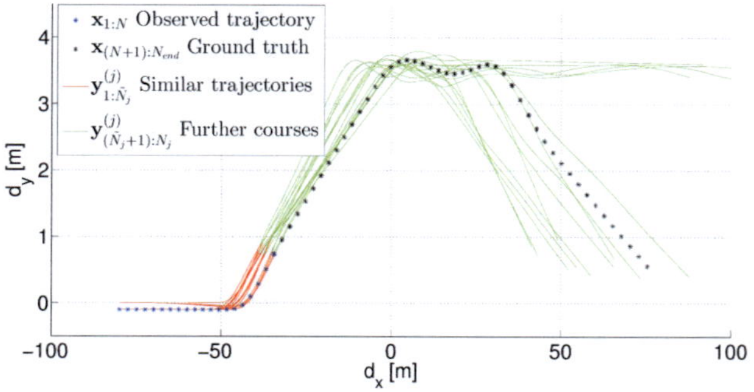

(c) $\epsilon = (5 \text{ m}, 0.25 \text{ m}, 10 \text{ km/h})$

Figure 5.5.: Retrieval of trajectories $\mathbf{y}$ of the database due to a given observation sequence $\mathbf{x}_{1:N}$ with different values for ϵ. Trajectories are chosen, if $d_{\text{lcs}}(\mathbf{x}_{1:N_x}, \mathbf{y}^{(j)}_{1:N_j}) < 1$ holds.

dicated as $\mathbf{y}^{(j)}_{1:\tilde{N}_y}$ and is received as the similar subsequence from the LCS computation. The further courses of these trajectories are indicated as $\mathbf{y}^{(j)}_{(\tilde{N}_y+1):N_y}$ and are used in the following to compute the required predictions for $\mathbf{x}_{1:N_x}$.

To **reuse and retain** the selected trajectories $\mathbf{y}^{(j)}_{(\tilde{N}_y+1):N_y}$ and to receive an adequate number of reasonable predictions to $\mathbf{x}_{1:N}$, different steps have to be conducted. The assumptions for the final set of predictions are that their number should be appropriate small and that the complete trajectory consisting of input sequence $\mathbf{x}$ and prediction should be passable (i.e., its patch has to be at least continuous). Furthermore, the final predictions should be associated with a probabilistic weighting value corresponding to Eq. (5.9). The following three steps are therefore performed sequentially:

- clustering all selected trajectories,
- merging the trajectory clusters to representatives,
- final transformation to receive drivable (continuous) trajectories.

In the following, $\mathbf{y}^{(j)}_{(\tilde{N}_y+1):N_y}$ is denoted simplified as $\mathbf{y}^{(j)}$. For clustering the trajectories different approaches may be applied. Algorithms like k-means result in a predefined number of clusters, which is not preferable in the context of this thesis. To provide a varying number of clusters, depending on the number and courses of the input trajectories $\mathbf{y}^{(j)}$, an agglomerative clustering algorithm is used in this thesis as shown in Algorithm 3. As a first step, every trajectory is interpreted as a singleton cluster. Afterwards, two clusters are combined iteratively depending on a minimization of the distance value between two clusters. Obviously, the LCS distance d_{lcs} of Eq. (5.7) is used applying the same threshold ϵ. The procedure terminates if

Algorithm 3 Clustering algorithm for trajectories

Require: clustering threshold $c_{clu} \in [0,1]$
Require: trajectories $\mathbf{y}^{(j)}$, building one cluster each
  for all $a_1, a_2 \in [0,\ldots,J], a \neq b$ **do**
    $d_{A_1,A_2} \leftarrow d_{\text{lcs}}(\mathbf{y}^{(a_1)}, \mathbf{y}^{(a_2)})$
  end for
  while $\exists$ Clusters A_1, A_2: $d_{A_1,A_2} < c_{clu}$ **do**
    assign clusters A_1, A_2 with $d_{A_1,A_2} = min!$ to single cluster
    update all distances: $d_{A_1,A_2} \leftarrow \frac{1}{|A_1||A_2|} \sum_{a_1 \in A_1} \sum_{a_2 \in A_2} d_{\text{lcs}}(\mathbf{y}^{(a_1)}, \mathbf{y}^{(a_2)})$
  end while

there are no more clusters A, B with a common distance value $d_{\text{lcs}}(A,B)$ smaller than a predefined cluster threshold c_{clu}.

In Figure 5.6 the results of the clustering algorithm are shown for the already known situation. The resulted clusters A_1, A_2, A_3 and A_4 are illustrated in different colors. As can be seen, one cluster represents trajectories of an overtaking maneuver with only one lane change. Three different kinds of overtaking maneuvers, varying mainly in the way of the second lane change, are represented by the other three clusters. By decreasing the threshold c_{clu} the number of clusters would be increased.

The result of the clustering algorithm is a number of K trajectory clusters, consisting of trajectories $\mathbf{y}^{(i_k)}$, with $k = 1,\ldots,K$ and $i_k = 1,\ldots,|A_k|$. Here, i_k is the trajectory index of the k-th cluster A_k and $|A_k|$ is the number of trajectories in this cluster. These clusters have to be merged into single representative trajectories each. Therefore, two different steps have to be executed:

First, a weighted mean $\hat{\mathbf{y}}^{(k)}$ for each cluster is computed, taking into account all trajectories of the cluster and their as-

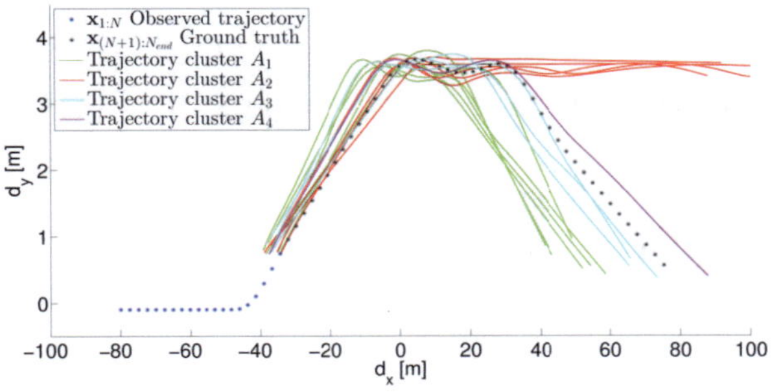

Figure 5.6.: Results of the agglomerative clustering of Algorithm 3. The stop criterion c_{clu} is set to 1.

sociated probabilities according to Eq. (5.9):

$$\hat{\mathbf{y}}^{(A_k)} = \sum_{i_k \in A_k} P(\mathbf{y}_{1:N_y}^{(i_k)} | \mathbf{x}_{1:N}, \lambda_{over}) \mathbf{y}^{(i_k)} \tag{5.10}$$

Thus, trajectories which are more similar to the input sequence $\mathbf{x}_{1:N}$ have a higher impact on the resulted, merged trajectory $\hat{\mathbf{y}}^{(A_k)}$ than trajectories of the same cluster which are less similar to $\mathbf{x}_{1:N}$ (*less similar* means a higher distance value d_{lcs}). Since the trajectories vary in their lengths, a normalization in time has to be conducted. In this thesis, the DTW algorithm is used for that purpose with respect to the longest trajectory $\mathbf{y}^{(i_k)}$. The result of this merging procedure is shown in Figure 5.7.

The second step is to calculate the combined probabilities for each cluster, which is easily accomplished by summing up over all trajectory probabilities of the cluster. As a result, each cluster A_k has a corresponding probability value $P(\hat{\mathbf{y}}^{(A_k)} | \mathbf{x}_{1:N}, \lambda_i)$.

The final step to receive the predictions $\hat{\mathbf{x}}^{(k)}$ (see Figure 5.4) is to align the end of the input sequence $\mathbf{x}_{1:N}$ (i.e., $\mathbf{x}_{1:N}(N)$) with the start of $\hat{\mathbf{y}}^{(A_k)}$ (i.e., $\hat{\mathbf{y}}^{(A_k)}(1)$). Therefore, a monotonous function $f^{(k)}$ is used to compensate the difference $z^{(k)} := \hat{\mathbf{y}}^{(A_k)}(1) -$

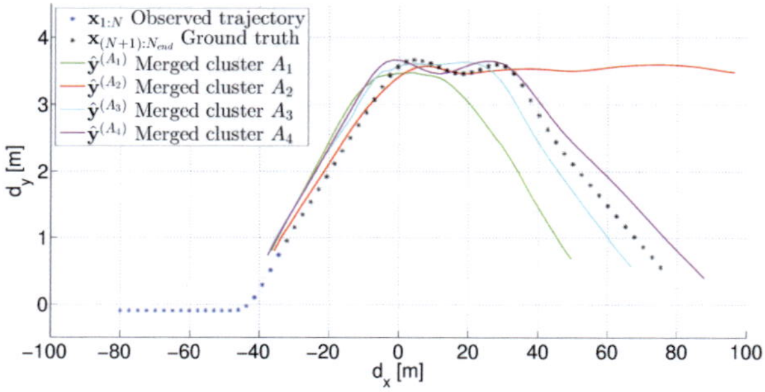

Figure 5.7.: Results of the merging procedure for all trajectory clusters according to Eq. (5.10).

$\mathbf{x}_{1:N}(N)$ resulting in a continuous trajectory $\hat{\mathbf{x}}^{(k)}$ in every dimension d

$$\hat{\mathbf{x}}^{(k)} = \hat{\mathbf{y}}^{(A_k)} + z^{(k)} f^{(k)}, \tag{5.11}$$

where $f^{(k)}$ is a monotonically decreasing function of the same length as $\hat{\mathbf{y}}^{(A_k)}$ with codomain $[0,1]$. Prediction results are shown in Figure 5.8, where f is used in terms of a linear function. The probabilities $P(\hat{\mathbf{x}}^{(k)}|\mathbf{x}_{1:N}, \lambda_{over})$ of every possible prediction are

$$P(\hat{\mathbf{x}}^{(1)}|\mathbf{x}_{1:N}, \lambda_{over}) = 0.46, \qquad P(\hat{\mathbf{x}}^{(2)}|\mathbf{x}_{1:N}, \lambda_{over}) = 0.28,$$
$$P(\hat{\mathbf{x}}^{(3)}|\mathbf{x}_{1:N}, \lambda_{over}) = 0.19, \qquad P(\hat{\mathbf{x}}^{(4)}|\mathbf{x}_{1:N}, \lambda_{over}) = 0.07,$$

which are the same as $P(\hat{\mathbf{y}}^{(A_k)}|\mathbf{x}_{1:N}, \lambda_{over})$.

Finally, the complete trajectory $\mathbf{x}$ has to be included to the database in the **retain** stage. Therefore, the trajectory of the complete maneuver $\mathbf{x}_{1:N_{end}}$ is recorded and saved as depicted in Figure 5.4. Thus, the end of the trajectory has to be defined which depends highly on the used sensor setup. In cases of limited sensor range and field of view the detection range of traffic participants defines implicitly maneuver's start and end. Car-to-Car messages, used as input information as presented in

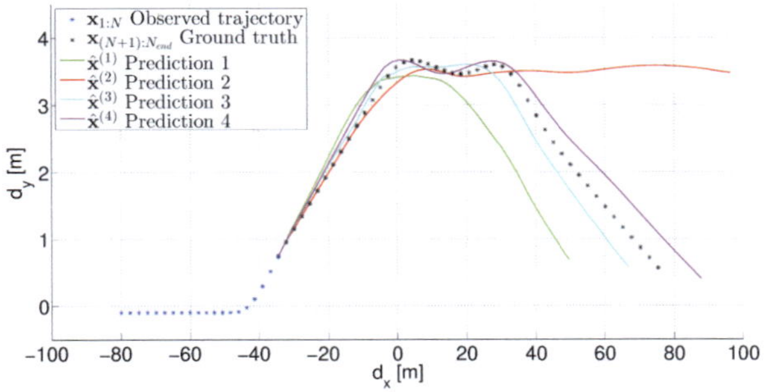

Figure 5.8.: Results of the prediction approach for a given input sequence and assuming an overtaking maneuver.

Chapter 4 require other approaches, like the usage of relative distances as the definition of the point in time of start and end of the maneuver.

5.3. Results and Evaluation

In this section, results of the proposed prediction approach with simulated data (Section 5.3.1) and with real world data (Section 5.3.2) are presented. Possible evaluations are presented afterwards in Section 5.3.3. For all results, the prediction thresholds are set to $c_{sim} = c_{clu} = 1$.

5.3.1. Results of Maneuver dependent Prediction

In the following, results for simulated sequences are presented using the same procedure as detailed in the last section. In all cases a specific maneuver is assumed to be executed, which corresponds to the exact driving behavior. The combination of these predictions with the probabilities of the maneuvers λ_i is presented in the next section using real world data.

Overtaking Maneuvers: In the last section one exemplary sequence of an overtaking maneuver was depicted consisting of two lane changes using $\epsilon = (5\text{m}, 0.25\text{m}, 10\text{km/h})$ as LCS threshold. Trajectory prediction was triggered at one characteristic point in time. In the following, results of another sequence performing an overtaking maneuver are presented, where prediction is triggered at different points in time. For more results this thesis refers to the appendix of (Omerbegovic, 2012).

In Figure 5.9 a sequence is evaluated assuming an overtaking maneuver. Prediction is triggered at four different points in time of the maneuver, resulting in different numbers K of trajectory predictions with different probabilities. For the first three triggered predictions of Figure 5.9(a), Figure 5.9(b) and Figure 5.9(c), one prediction represents an overtaking maneuver with only one lane change (trajectory prediction $\hat{\mathbf{x}}^{(2)}$ in all figures). The corresponding probability $P(\hat{\mathbf{x}}^{(2)}|\mathbf{x}_{1:N}, \lambda_{over})$ decreases when going forward in the sequence, corresponding to the ground truth of the further course of the trajectory. In other words, the later the prediction is triggered, the greater the probability of correctly predicting a maneuver with two lane changes is. At the last point in time of Figure 5.9(d), the second lane change is currently executed. Thus, only one trajectory is predicted, which is clustered from all retrieved trajectories. Consequently, the calculated probability is $P(\hat{\mathbf{x}}^{(1)}|\mathbf{x}_{1:N}, \lambda_{over}) = 1$ in this case.

Following Maneuvers: For trajectory prediction assuming a following maneuver, the same approach is used as outlined in the last section for overtaking maneuvers. As the dynamic complexity is not comparable to overtaking maneuvers, conventional prediction approaches may be also applied here. To enable a consistent framework, the proposed CBR-based approach is used instead. The used database of Figure 3.5 only consists of about 100 sequences (in comparison with about 900

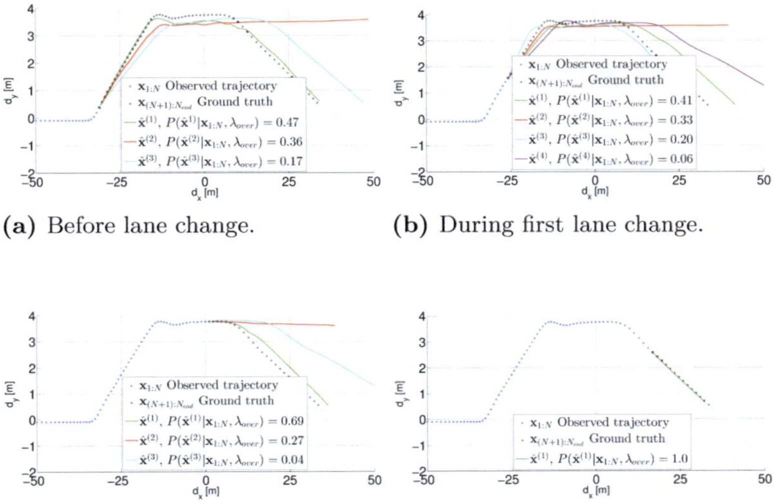

(a) Before lane change. **(b)** During first lane change.

(c) During passing of the vehicles. **(d)** During second lane change.

Figure 5.9.: Trajectory prediction results for one sequence at different points in time assuming an overtaking maneuver including their probabilities $P(\hat{\mathbf{x}}^{(i)}|\mathbf{x}_{1:N}, \lambda_{over})$.

sequences of overtaking maneuvers), mainly differing in the relative distance of both vehicles. This safety distance is mostly depending on the absolute velocity of both vehicles. In the simplest case, vehicles are driving with the same constant velocity. Nevertheless, the used database also contains maneuvers with acceleration and deceleration stages as well as with rather small differences in the absolute velocities of both vehicles. The considered dimensions for the complete CBR cycle are again, d_x, d_y and v_{rel}, but additionally the absolute velocity of the heading vehicle v_{lead} is taken into account. Evaluation with both state vectors showed the better adaption on the scenario with the extended version. The threshold of the LCS computation is set to $\epsilon = (5\text{ m}, 1.75\text{ m}, 10\text{ km/h}, 10\text{ km/h})$, where the value of d_y is increased significantly.

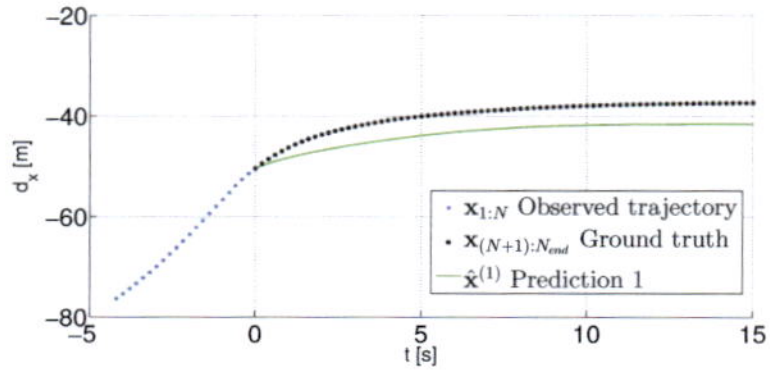

(a) Relative long. distance d_x.

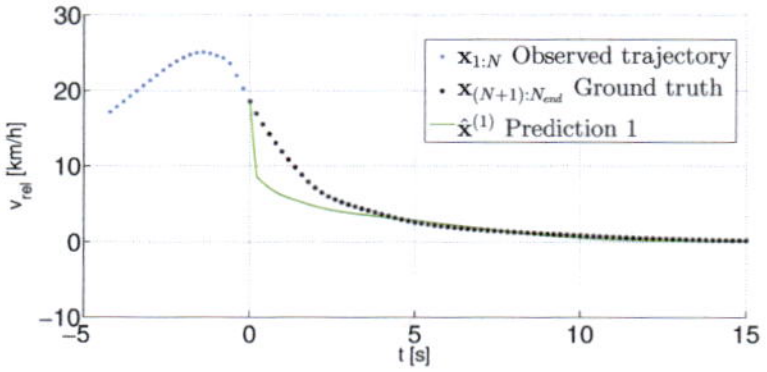

(b) Relative velocity v_{rel}.

Figure 5.10.: Trajectory prediction results for one sequence assuming a following maneuver.

Thus, also aborted overtaking maneuvers may be taken into account.[2]

Prediction results for one sequence describing a following maneuver are depicted in Figure 5.10. As for all other testing sequences, the prediction approach using the following database results in only one prediction $\hat{\mathbf{x}}^{(1)}$ (i.e., $J = K = 1$), with probability $P(\hat{\mathbf{x}}^{(1)}|\mathbf{x}_{1:N}, \lambda_{foll}) = 1$. For illustration issues, the relative lateral distance d_y and the relative velocity v_{rel} are plotted over time. The predicted trajectory describes a following maneuver with a safety distance of about 40 m, which corresponds well with the ground truth. In situations where the following distance does not corresponds to the safety distance of the current absolute velocity (e.g., performed by non-safety-conscious drivers), the prediction approach will perform worse while using the given database. However, prediction performance will be increased when the CBR system is adapted over time by retaining sequences of the same driver (or driver behavior) to the repository. The relative velocity tends to zero (as certainly the ground truth).

[2]As there is no general definition of the considered traffic maneuver, especially of their starting and ending points in time, other values of ϵ may also be applied.

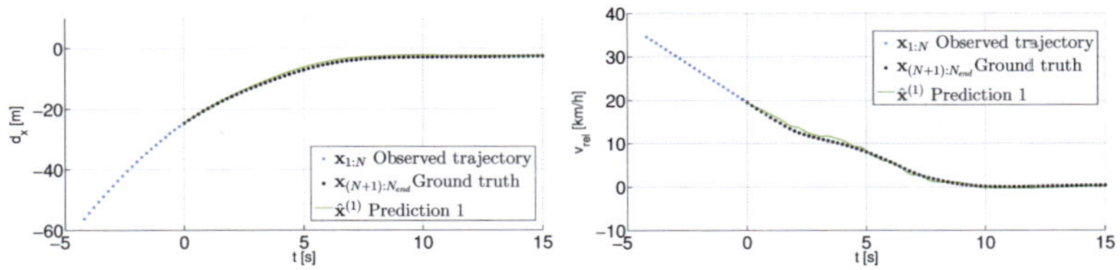

(a) Relative long. distance d_x.

(b) Relative velocity v_{rel}.

Figure 5.11.: Trajectory prediction results for one sequence assuming an flanking maneuver.

Flanking Maneuvers: Trajectory prediction for flanking maneuvers has similar properties as for situations assuming following maneuvers. The dynamic complexity is on a similar level, which motivates the small required database of about 50 sequences (see Figure 3.5). Furthermore, the resulted predictions also consists of only one trajectory $\hat{\mathbf{x}}^{(1)}$ with probability $P(\hat{\mathbf{x}}^{(1)}|\mathbf{x}_{1:N}, \lambda_{flank}) = 1$. The considered dimensions are d_x, d_y and v_{rel} as the absolute velocity of both vehicles does not have any impact. The threshold ϵ is set again to $\epsilon = (5\text{ m}, 1.75\text{ m}, 10\text{ km/h})$.

Prediction results for one sequence describing an approaching vehicle with a small relative velocity on the adjacent lane, decelerating and flanking the other vehicle are depicted in Figure 5.11. As they are the most crucial parameters, values for d_x and v_{rel} are plotted over time, again. The retrieve stage results in a set of ten similar trajectories ($J = 10$), which are all clustered to one single cluster ($K = 1$). As the execution of any flanking maneuver does not vary much in its relative kinematic dimensions, the prediction $\hat{\mathbf{x}}^{(1)}$ matches very well the ground truth of the trajectory $\mathbf{x}$.

5.3.2. Results of Prediction Approach

After showing different prediction results assuming different maneuvers, the contribution of this section is twofold:

- providing results using maneuver probabilities,
- providing results with real world data.

The first point corresponds to the final prediction results of the proposed approach, which are stochastically independent of the maneuver λ_i (see Figure 5.3).

For every query of the trajectory prediction, the probabilities $P(\lambda_i|\mathbf{x}_{1:N})$ of the maneuver recognition of Chapter 3 are computed. These values are used for weighting the predictions $\hat{\mathbf{x}}^{(k)}$ for each assumed maneuver. As the recognition results do not allow a meaningful interpretation, even in a logarithmic representation, a proper transformation to the range $[0, 1]$ has to be applied (as mentioned in Section 3.3, the used forward variables have usually very small values, which are individually hard to interpret).

A typical situation, where the usage of a probabilistic consideration of different traffic maneuvers is required, is presented in the following: A vehicle is approaching another traffic participant on the same lane, either accelerating or decelerating, and possible further trajectories may include overtaking maneuvers with single or multiple lane changes or even no lane change and overtaking maneuver at all. Single trajectory prediction, assuming each, a following and an overtaking maneuver are shown in Figure 5.12 (flanking maneuvers are neglected). In Figure 5.12(a) and Figure 5.12(b) the prediction results assuming an overtaking and a following maneuver are depicted, for the longitudinal and lateral relative distances d_x and d_y, and for the longitudinal distance d_x over the time t, respectively. Assuming an overtaking maneuver, there are two predictions

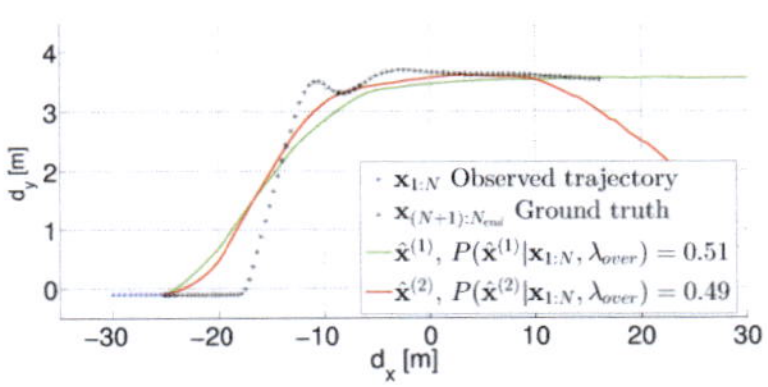

(a) Assuming an overtaking maneuver.
Parameters: d_y, d_x

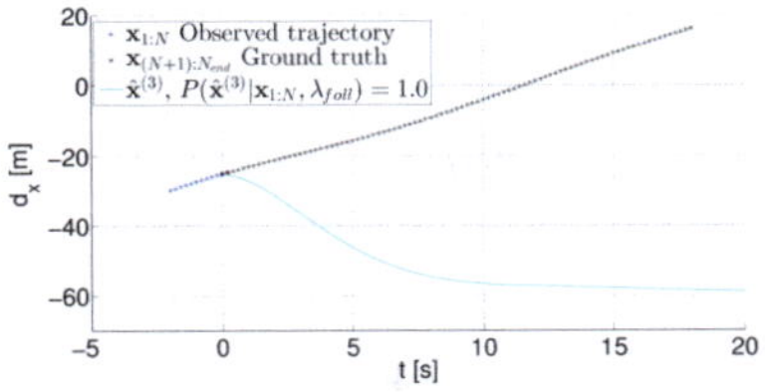

(b) Assuming a following maneuver.
Parameters: d_x, time t

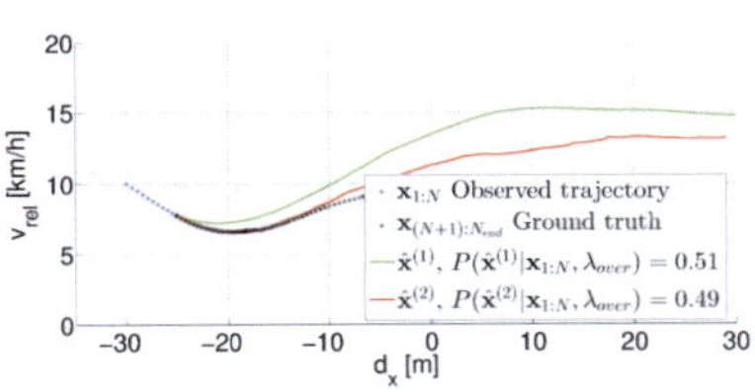

(c) Assuming an overtaking maneuver.
Parameters: v_{rel}, d_x

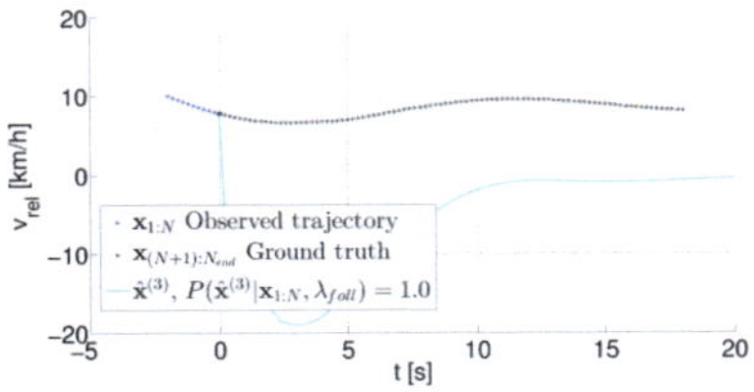

(d) Assuming a following maneuver.
Parameters: v_{rel}, time t

Figure 5.12.: Trajectory prediction results assuming each, a following and a flanking maneuver in terms of different parameters including their probabilities $P(\hat{\mathbf{x}}^{(i)}|\mathbf{x}_{1:N}, \lambda_i)$.

$\hat{\mathbf{x}}^{(1)}$ and $\hat{\mathbf{x}}^{(2)}$ describing a maneuver with one and two lane changes. Their probabilities are similar, which seems to be reasonable, as it is a very early point in time of the maneuver. As mentioned in the last section, the prediction approach assuming a following maneuver results in only one trajectory $\hat{\mathbf{x}}^{(3)}$ with probability $P(\hat{\mathbf{x}}^{(3)}|\mathbf{x}_{1:N}, \lambda_{foll}) = 1$ representing a specific safety distance (depending on the absolute velocity). In Figure 5.12(c) and Figure 5.12(d) the relative velocities are plotted.

Afterwards, the maneuver probabilities are used as weighting factors for all predicted trajectories. Respective results are shown in Table 5.1. For the probabilities of each maneuver

$\hat{\mathbf{x}}^{(k)}$	$P(\hat{\mathbf{x}}^{(k)}\vert\mathbf{x}_{1:N},\lambda_i)$	$P(\lambda_i\vert\mathbf{x}_{1:N})$	$P(\hat{\mathbf{x}}^{(k)}\vert\mathbf{x}_{1:N})$
$\hat{\mathbf{x}}^{(1)}$	0.51	0.59	0.30
$\hat{\mathbf{x}}^{(2)}$	0.49	0.59	0.29
$\hat{\mathbf{x}}^{(3)}$	1.0	0.83	0.83

Table 5.1.: Probabilities of the trajectory prediction approach of Figure 5.12 considering overtaking and following maneuvers ($k = 1, 2 \rightarrow \lambda_{over}$, $k = 3 \rightarrow \lambda_{foll}$).

holds: $P(\lambda_{foll}|\mathbf{x}_{1:N}) > P(\lambda_{over}|\mathbf{x}_{1:N})$. That is due to the deceleration during the input sequence $\mathbf{x}_{1:N}$, which is not an indicator for an upcoming overtaking maneuver. Thus, the prediction of a following maneuver is most likely with $P(\hat{\mathbf{x}}^{(3)}|\mathbf{x}_{1:N}) = 0.83$. Although, this does not match to ground truth of the depicted trajectory $\mathbf{x}$, it's seems to be reasonable, as the upcoming lane change maneuver is not obvious at this point in time. For more results at different points in time and for the combination of other maneuvers, please refer to (Omerbegovic and Firl, 2013) and (Omerbegovic, 2012).

An evaluation with real world data has to be done, to show the general applicability of the proposed approach. Sensor noise and the robustness to outliers in the input sequence are only two possible challenges, which cannot be handled with simulated data appropriately. Therefore, the sensor setup depicted in Figure 3.12 is used, where a fusion of camera and radar data is applied.

In Figure 5.13, prediction results for one sequence are depicted at different points in time. Firstly, the noisy sensor data has to be considered, mainly resulted due to the bad lateral resolution of the used radar sensor. To be aware of that, the LCS threshold is increased in its lateral distance component and is set to $\epsilon = (5\text{ m}, 1\text{ m}, 10\text{ km/h})$. The different stages of the se-

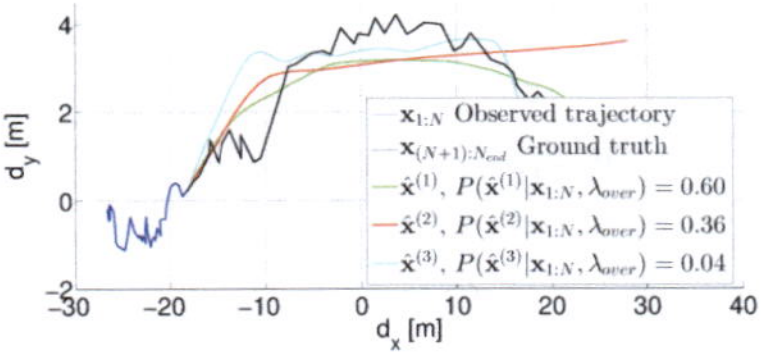

(a) Before lane change.

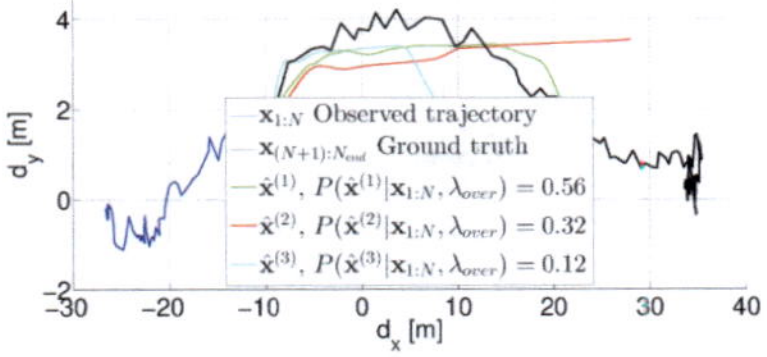

(b) During first lane change.

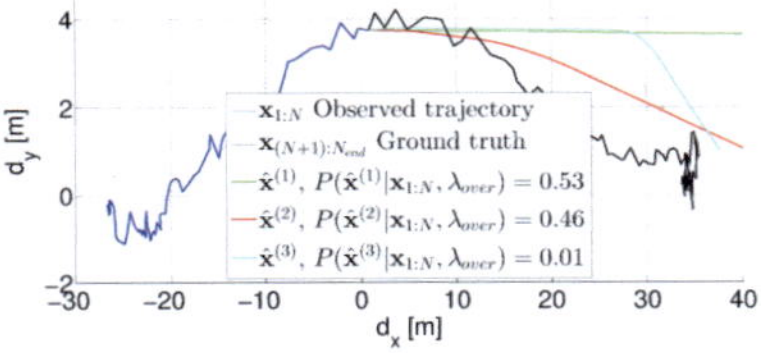

(c) During passing of the vehicles.

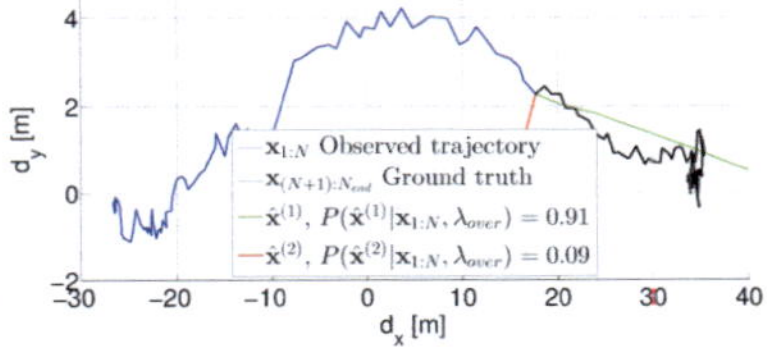

(d) During second lane change.

Figure 5.13.: Trajectory prediction for one sequence captured by the sensor setup of Figure 3.12 at four different points in time assuming an overtaking maneuver including their probabilities $P(\hat{\mathbf{x}}^{(i)}|\mathbf{x}_{1:N}, \lambda_{over})$.

quence are similar to the one evaluated in Figure 3.14, as one vehicle is firstly following another one, suddenly performs an overtaking maneuver including two lane changes. For the first three predictions (Figure 5.13(a) - Figure 5.13(c)) two different possible trajectories with a second lane change and one trajectory indicating an overtaking maneuver with only one lane change are predicted. At the last prediction of Figure 5.13(d), where the second lane change is being executed, no more trajectories keeping the left lane can be retained from the repository correctly, and thus no trajectory cluster k is built. All predictions represent reliable trajectories for the given scenario, although their probabilities seem to be hardly interpretable. This is an expected phenomenon for such situations, where also human experts have problems in performing accurate prediction (e.g., one lane change vs. two lane changes)[3].

Nevertheless, the prediction results become more accurate, when being combined it with the corresponding maneuver probabilities $P(\lambda_i|\mathbf{x}_{1:N})$. Their results are mapped into the range $[0,1]$ by applying a monotonic transformation. As the maneuver recognition results are usually given by their natural logarithm (see figures of Chapter 3), the two required boundaries of the mapping function are $c_{low} < 0$ mapped to the probability $P(\lambda_i|\mathbf{x}_{1:N}) = 0$ and $c_{high} = 0$ mapped to the probability $P(\lambda_i|\mathbf{x}_{1:N}) = 1$ (as $\ln(0) = 1$). Another possible transformation, especially when considering the probabilities of two different maneuvers λ_1 and λ_2, is to analyze their ratio (see Eq. (4.1)) and map the single probabilities to $[0,1]$ by remaining their ratio value.

For the sequence evaluated in this section, the results of the maneuver recognition at the four different points in time are depicted in Table 5.2. As already mentioned, trajectory prediction assuming a following or a flanking maneuver results in only one prediction each (which are not illustrated separately in Figure 5.13). The final predictions indicating an overtaking trajectory are combined regarding the number of performed lane changes resulting in four different predictions. Probabilities for following and flanking trajectories are very well interpretable, as e.g., the following probability $P_{\hat{\mathbf{x}},foll}$ is equal to 0, when the two vehicles performs the passing stage of the overtaking maneuver. When currently starting the first lane change, a rather low probability of 0.25 for a following trajectory represents the reliable uncertainty, if the lane change is executed or the maneuver is stopped. Both overtaking probabilities ($P_{\hat{\mathbf{x}},over1}$ and $P_{\hat{\mathbf{x}},over2}$) are nonzero, expect for the last prediction of Figure

[3]In fact, also the driver himself is not always being sure of how his own maneuver will continue when starting it. Thus, a reliable prediction of the trajectory in a probabilistic manner is rather impossible in such situations.

	P_{over}	P_{foll}	P_{flank}	$P_{\hat{\mathbf{x}},over1}$	$P_{\hat{\mathbf{x}},over2}$	$P_{\hat{\mathbf{x}},foll}$	$P_{\hat{\mathbf{x}},flank}$
Fig. (a)	0.75	0.25	0.0	0.27	0.48	0.25	0.0
Fig. (b)	0.78	0.18	0.04	0.25	0.53	0.18	0.04
Fig. (c)	0.86	0.0	0.14	0.46	0.40	0.0	0.14
Fig. (d)	0.78	0.20	0.02	0.78	0.0	0.20	0.02

Table 5.2.: Maneuver probabilities at the four distinct points in time of trajectory prediction, see Figure 5.13. The transformed maneuver probabilities are $P_{over}, P_{foll}, P_{flank}$, which are each mapped to the range $[0, 1]$, with $\sum_i P_i = 1$. Additionally, the probabilities of the trajectory indicating a following ($P_{\hat{\mathbf{x}},foll} := P(\hat{\mathbf{x}}^{(foll)}|\mathbf{x}_{1:N})$) or flanking ($P_{\hat{\mathbf{x}},flank}$) maneuver or an overtaking with one ($P_{\hat{\mathbf{x}},over1}$) or two lane changes ($P_{\hat{\mathbf{x}},over2}$) are depicted.

5.13(d), where the second lane change has already started, and consequently $P_{\hat{\mathbf{x}},over1} = 0$ holds.

5.3.3. Evaluation

After showing and analyzing results of the proposed prediction approach, in this section the framework is evaluated by comparing its results to ground truth data, as well as with a classical predictor, namely a Kalman filter. A database has to be defined for both of them, including trajectories for which prediction is triggered at different, predefined points in time. The set of used trajectories includes sequences of overtaking maneuvers, as already used for the CBR repository in the last section, extended to stages of following (at the beginning of the sequence) and flanking (in the middle of the sequence). Trajectory prediction is triggered for all sequences (excluded of the CBR database for this case).

Comparison with ground truth: For a comparison of the prediction results with ground truth data, a reasonable definition of a prediction error has to be done. Assuming the point in time of prediction as N, the prediction horizon as ΔN and K

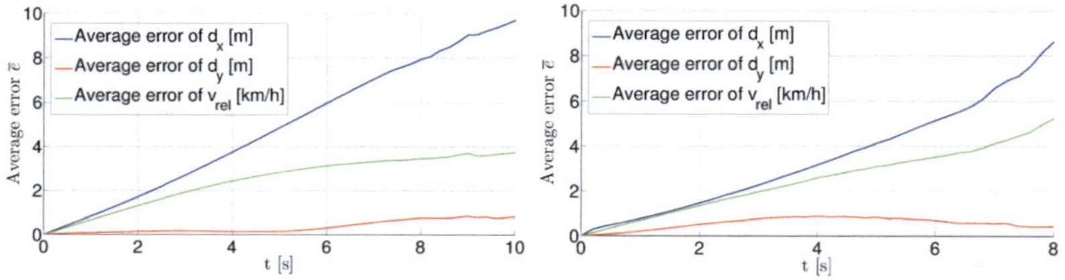

(a) Before lane change. (b) During passing of the vehicles.

Figure 5.14.: Evaluation of the proposed trajectory prediction approach in comparison to ground truth data for d_x, d_y and v_{rel} at two different points in time of prediction.

predictions for a given input sequence $\mathbf{x}_{1:N}$. Then, the overall prediction error for this sequence is given as:

$$e = \sum_{k=1}^{K} P(\hat{\mathbf{x}}^{(k)}_{N:(N+\Delta N)} | \mathbf{x}_{1:N}) |\hat{\mathbf{x}}^{(k)}_{N:(N+\Delta N)} - \mathbf{x}_{N:(N+\Delta N}|, \quad (5.12)$$

where the errors for each single prediction are weighted due to their calculated probabilities. e is computed for each dimension (d_x, d_y, v_{rel}) separately. For a given number of Q trajectories, for which prediction is queried, the overall error of the proposed approach is given as:

$$\overline{e} = \frac{1}{Q} \sum_{q=1}^{Q} e_q, \quad (5.13)$$

where e_q is the prediction error of the q-th trajectory according to Eq. (5.12).

In Figure 5.14 the prediction errors for the relevant dimensions are shown for a time horizon of 10 and 8 seconds, respectively. The smaller prediction horizon in Figure 5.14(b) is due to the limited length of some trajectories in the database. The functions $\overline{e}(t)$ are mainly increasing, especially for d_y and v_{rel}.

Even the prediction error for the longitudinal distance d_x are reaching up to 10 meters, this result is at least acceptable when taking the large prediction range of 10 seconds into account. Regarding the results for the lateral distance d_y, the error remains quite small, even for large prediction time horizons. For both examples, the error $\overline{e}$ for d_y is below 1 meter, which has to be noticed as very convincing, as the used sequences includes a plenty of lateral, dynamic maneuvers. These are usually one of the most challenging situations for trajectory prediction approaches.

Comparison with KF prediction: The comparison of the prediction approach with ground truth data does not allow a fully representative evaluation of the proposed predictor. Therefore, the comparison to state of the art trajectory prediction approaches has to be drawn. In the following, a Kalman filter is used as one of the most common techniques. Two different assumptions are made, for vehicle's motion in longitudinal and lateral direction:

- longitudinal motion is according to the discrete Wiener process acceleration (DWPA) model.
- lateral motion is according to the discrete white noise acceleration (DWNA) model.

Details of both model assumptions can be found in (Bar-Shalom et al., 2001) and are not discussed in this thesis.

For a comparison of both prediction methods, the same database is adopted, as used in the previous paragraph. Therefore, the error e is calculated at 4 different points in time of the sequences (see Figure 5.13). For a better comparison of both prediction methods, the prediction horizon is limited to 3 seconds. In Figure 5.15 the results for the dimensions d_x, d_y and v_{rel} are depicted. For the longitudinal, relative distance

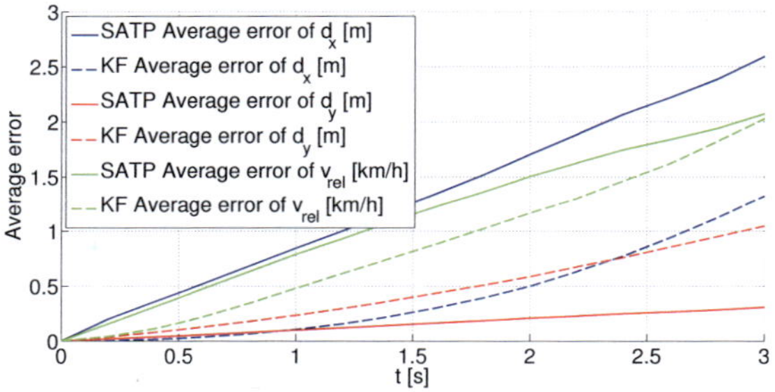

Figure 5.15.: Comparison of the proposed situation adaptive trajectory prediction (SATP) approach and a Kalman filter for d_x, d_y and v_{rel}.

d_x, the proposed adaptive prediction method performs worse than the KF predictor. This originates from the underlying motion model, which approximate the driving behavior accurately in most cases. Even if the course of the driving maneuver is not predicted correctly (e.g., the number of performed lane changes), the error of d_x keeps often comparative small. For the predicted relative velocity v_{rel}, the error of the KF is again smaller than the one of the maneuver dependent prediction approach. However, the difference between both prediction approaches is much smaller, and actually would change its sign if observing larger prediction horizons. The main advantage of the approach of this thesis is reflected best in the error curves of the lateral distance d_y. As the KF is not able to predict vehicle's trajectories accurate when performing maneuvers including multiple lane changes, the corresponding error is quite large. Furthermore, the difference between both prediction errors would increase if taking larger prediction horizons into account. The reason therefore is the simplified assumptions of the acquired motion model in comparison of the flexible CBR-based approach of this thesis.

5.4. Summary

In this section an approach was presented how the maneuver recognition results may be used for predicting vehicle's trajectories in a probabilistic way. Thus, the probabilistic information on the execution of different interactions between traffic participants is used to robustly predict their future courses on a largest possible time horizon. Therefore, two different steps are executed. Firstly, trajectories are predicted for all relevant maneuvers assuming the specific interaction is true. The second stage combines the resulting prediction for every maneuver with the corresponding probabilities according to the procedure of Chapter 3.

The second step implies some simplified multiplications only as a weighting, the focus of this chapter was on how to predict a trajectory assuming a single maneuver as given. The conceptual design follows the principle of Case-based reasoning (CBR), as for each maneuver a database is constructed. When prediction is queried, the database is searched for *similar* trajectories (*cases*). These trajectories are used to draw conclusions to possible predictions of the further course of the current trajectory (included in the *reuse* and *revise* CBR stage). The applied distance measure is the Longest Common Subsequence (LCS). The complete prediction framework, including the *retain* stage, is able to adapt on different drivers or driving characteristics.

Results are gathered with simulated as well as with real world data. Different types of sequences were tested, where prediction was triggered at different points in time. Furthermore, single results for one assumed maneuver were presented, as well as the final combined results using the corresponding maneuver probabilities. For an evaluation of the proposed approach, a comparison with ground truth data was outlined. Furthermore, a state of the art Kalman filter was used for benchmarking the system of this thesis.

6. Conclusion

To account for the increased functionality of future advanced driver assistance systems (ADASs), the understanding of more and more complex situations becomes requisite. Therefore, extended modeling approaches have to be applied to recognize and consider all crucial properties in the traffic scene. In this work a framework was proposed to model and assess extra urban traffic situations. This scenario information may assist a variety of different features. After reviewing different approaches, a probabilistic model was selected for recognizing maneuvers of multiple traffic participants, independent of the used sensor setup and of the triggered ADAS feature.

The basic concept to model and recognize traffic maneuvers was outlined in Chapter 3. Therefore, the exact definition of the problem was given first, considering maneuvers in terms of interactions between multiple traffic participants with spatio-temporal dependencies (overtaking, following and flanking maneuvers). Hidden Markov Models (HMMs) were chosen for that purpose, which fit best the given requirements of this thesis due to their structure and efficient training and evaluation algorithms. At the beginning, the input data has to be defined properly, as it directly corresponds to the recognition capabilities of the model. Basically, the relative kinematic information between the interacting vehicles was considered therefore. To take more characteristic information of the traffic situation into account, additional a priori knowledge (e.g., about different road types) was added using Bayes' theorem. For the challenging consideration of multiple other objects the required free space for the execution of the maneuver was added to the model's observation vector. Thus, occupancy grids were introduced in this thesis, adapted on the current scenario. Therewith static objects may be considered, like road boundaries or construction sites, as well as other dynamic traffic participants and their predicted trajectories. This flexible approach enabled a very accurate modeling of the entire traffic situation by keeping model complexity low at the same time. Model complexity is thus independent of the number of traffic participants in the scene which have to be considered. Only the calculation of the occupancy value becomes more challenging with an increased number of vehicles.

The recognition stage of the proposed framework was executed by using the forward algorithm, which computes effectively the conditional probability of the observation sequence for all given maneuver models for one point in time. These models had to be trained before using simulated data and taking different driver and vehicle characteristics into account. Recognition results were presented with simulated as well as with real world data for different types of sequences of extra urban traffic scenarios. They were visually analyzed by a comparison to human expert knowledge, showing the general reliability or the approach. Also for input data containing heavy noise, the calculated maneuver probabilities can be still interpreted well.

Even if the proposed model enables the consideration of multiple traffic scenario characteristics (and thus enables realistic recognition results in most situations), there are still some open issues which are not handled so far. All scenarios discussed in this thesis are assumed to consist of a constant number of lanes, which may vary in their occupied status. Thus, scenarios like access roads or motorway exits will results in non-plausible recognition results, due to the non-consideration of a priori knowledge about the driving behavior of traffic participants in such scenarios. Another drawback of the presented modeling concept is the limited prediction capabilities due to the missing of an underlying kinematic motion model of each traffic participant. One open task of future work is the integration of motion assumptions into the maneuver modeling framework. As this is hardly possible with the proposed HMM, other approaches have to be discussed here. One possible solution may be the usage of factorial HMMs. Besides the disadvantageous need of approximate inference algorithms, factorial HMMs are able to add structural information about the underlying system, what includes enhanced predictive modeling capabilities (see also (Ghahramani and Jordan, 1997)).

One of the main challenges for all situation assessment approaches was discussed in Chapter 4: their evaluation. As it highly depends on the context of addressed ADAS feature, no general evaluation method is available. In this thesis, a method was presented by predicting safety critical situations, which are crucial for most systems. Thus, dynamic lane change maneuvers and abrupt brakings were selected as occurring in many extra urban scenarios. The prediction of these situations was realized by analyzing the ratio of two different maneuver probabilities each. Evaluation was done using a database of more than 100 sequences of real world data for both situations, focusing on two different aspects: As a first step, the prediction time with respect to the point in time of the real maneuver execution was analyzed. Since many features require a general robustness to wrongly triggered executions (as for most driver warning and braking or steering intervention systems), an evaluation of the ratio of true and false positive rates was performed afterwards. Thereby, the high potential of the presented method was pointed out, by predicting most safety critical situations at an early point in time with a reasonable low false positive rate.

The benefit for one example ADAS feature was outlined in Section 4.2, where also the independence of the used sensor setup was demonstrated by applying Car-to-Car communication (C2C) messages as input data. The triggered feature in this work was a security module which verifies the trustworthiness of incoming C2C messages and was used in addition to other security approaches like cryptography. The basic concept of message verification was based on a deviation of received mobility information in the message (e.g., position, speed) with a prediction based on already received messages and an adequately selected motion model. The Kalman filter which was used for this purpose turned out to be an effective estimator in most situations. However, significant flaws have been identified in cases of highly dynamic traffic scenarios, where the verification flow was adapted by the proposed probabilistic maneuver prediction. An adaption of the Kalman gain was introduced, when correctly predicting a dynamic lane change or abrupt braking maneuver, leading to a significantly enhanced accuracy and robustness of the overall verification system. The complete framework was implemented and integrating into an already existing software architecture of the sim^{TD} project.

By applying the probabilistic maneuver recognition framework to the domain of C2C communication and its security aspects two different attainments have to be pointed out. Firstly, the usability of the C2C technology as input data of the recognition approach of this work was shown, including its evaluation yielding promising results. Secondly, the integration of the proposed concept into a complete software architecture proves the possible benefit for ADAS features. However, there are still some aspects, which have not been analyzed completely so far. Up to now, only the motion information included in Cooperative Awareness Messages (CAMs) were considered for message verification. Especially during high dynamic situations, event based messages as Decentralized Notification Messages

(DENMs) may be included to increase the reliability of the proposed framework. Examples therefore are the information about a lane narrowing as at the beginning of a construction zone, or the information on certain accident events as skidding vehicles. Considering this information will increase system's robustness by adapting the a priori probabilities of certain maneuvers in those cases and evaluate incoming messages with respect to that knowledge. Furthermore, system's behavior should be analyzed, when being in situations like acceleration lanes or interchanges, as they are not directly considered in the used maneuver modeling.

In Chapter 5 another benefit of the calculated probabilistic information about different traffic maneuvers was pointed out. The prediction of future trajectories of single traffic participants was analyzed with respect to the recognition results. As most state of the art prediction approaches use a specific motion model, optionally adapted using information on road geometry or assuming collision-free trajectories, those lack in robust predictions in dynamic traffic maneuvers for large time horizons. As reasonable predictions using the presented HMM based method are not possible, its probabilistic information was integrated into a prediction framework motivated by Case-based reasoning (CBR). Therefore, databases of trajectories for each considered maneuver were built up by varying different characteristic parameters of the maneuver execution. Prediction for a given sequence was done by first retrieving similar trajectories (cases) of the databases, which was done by using an adaption of the Longest Common Subsequence (LCS). This similarity measure was mapped to a probability value for each trajectory which was passed through the further process. These cases were used in the reuse and revise stage, to infer about the further course for the given trajectory. Closing the CBR cycle with the retain stage offers the prediction framework the possibility to be adapted to different driver and driving characteristics. The last prediction step combined the resulted predictions with the recognition results for each maneuver yielding in multiple possible trajectories with associated probabilities.

Results for the presented prediction approach were presented using simulated and real world data. Therefore, different sequences of maneuvering vehicles were observed proving the plausibility of the resulted predictions at different points in time. Predictions assuming one given maneuver were computed as well as the final results, combined with the corresponding maneuver probabilities. Evaluation was performed by comparing the predicted trajectories including their probabilistic information with ground truth data as well as with a state of the art predictor, where a Kalman filter was used in this thesis.

The presented trajectory prediction approach offers some major advantages compared to commonly used predictors. As motivated by the shortcomings of most predictors when being executed in dynamic traffic maneuvers, the proposed framework has its strength in those situations. There, prediction results are accurate, even when calculated over large time horizons of some seconds. Furthermore, the probabilistic nature enables prediction results which match well with the interpretation of human experts concerning those situations. Different further courses of single vehicles are usually possible and handled in this work, especially when multiple traffic participants are involved.

However, there are still some drawbacks, which will motivate some future work in this field. The comparison with the KF indicates that the missing motion model, causes larger prediction errors particular in dimensions of longitudinal position and velocity. Thus, a combination of the proposed concept with an underlying motion model would help to overcome this shortcoming. One possible solution for this may be to add another step to the framework, where the received predictions are adapted by using some predefined motion assumptions. Furthermore, the implementation in this work requires both, memory and processing power as the used databases are appropriate large to model all variants of the maneuver execution. To overcome this problem, the overall size of the database has to be limited, e.g., by deleting non-used or *too* similar cases. However, the impact on the prediction accuracy has to be evaluated therefore.

Furthermore, to close the loop from trajectory prediction to the proposed maneuver recognition framework, a usage of this prediction approach to model the impact of dynamic traffic participants on the execution and required free space of traffic maneuvers is desirable. Therefore, the used sensor systems of this thesis are not able to provide an adequate quality of kinematic information about relevant objects with respect to detection range, field of view and a high position accuracy.

A. Appendix

A.1. Sensor and Software Architecture for Observing Traffic Maneuvers

In Chapter 3, an experimental vehicle was described gathering real world data for evaluating the proposed maneuver recognition approach. In the following the used sensor setup is described, including the applied development environment.

The vehicle's perception system was built up at Adam Opel AG while participating in the German research initiative *Aktiv*, more precise in the containing subproject *Active Safety - Active Hazard Braking*, see the project's web page (Aktiv, 2010) for more information. The vehicle was used as an observer of further vehicles performing certain traffic maneuvers in the forward sensor field of view (see Figure 3.12). Radar and camera image data are both processed on the vehicle's hardware platform, basically consisting of 5 Car-PCs for sensor data fusion and situation analysis tasks.

The fusion of sensor input data from the ARS300 radar sensor and the CSF200 mono camera is mainly done by the following principles. For detection purposes of traffic participants or other objects, data perceived by the radar sensor is processed mainly. For classifying the received objects, information of radar and camera are both taken into account. While camera images provide quite dense scenario information, the radar sensor data makes kinematic object properties available directly. Both advantages are employed extensively for classifying objects under different environment conditions. In many cases, especially when detecting an object for the first time, camera image data is used for verifying the objects given from the radar sensor. When objects are detected and classified, they are tracked over time by both sensors, depending on the conditions for detecting them at different points in time (weather conditions, line of sight and separation of nearby objects). So the advantages of both sensors are combined.

Besides these steps for detecting, classifying and tracking objects, an interpretation of the current scene was implemented for a best possible execution of the hazard braking functionality. Therefore, the traffic situation is analyzed with respect to different situation types of rear-end collisions. For that purpose all detected road lanes are marked with a flag indicating its occupancy status. A lane is marked as *free* if no other dynamic objects or road boundaries are detected, otherwise it is *occupied* (compare the continuous valued definition of the occupancy grids of this thesis in Section 3.2.4). Depending on the number of lanes and its occupancy status different braking and warning interventions are triggered if an upcoming collision is recognized.

For evaluating the complete system of recognizing and tracking other traffic participants including an interpretation of the scene regarding the automated hazard braking system, an online evaluation environment is used, as depicted in Figure A.1. The interface is spitted into two parts: On the right side the camera image is depicted including all radar detections (yellow rectangles) and all classified objects (red rectangles). Below this image basic buttons for recording the complete data stream are located. On the left side basic information of the situation analysis is shown. Therefore, a virtual top view of the traffic scene is visualized containing the information of all detected road boundaries and lane markings. Furthermore, all classified vehicles are displayed whose colors are depending on weather they are relevant braking targets are not.

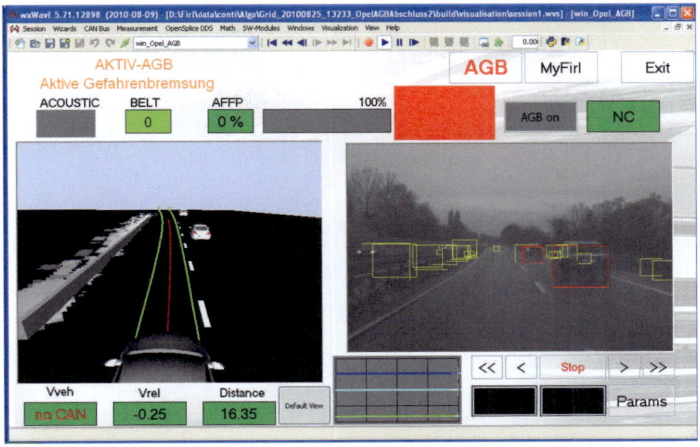

Figure A.1.: Development Environment used for extracting and analyzing data for maneuver recognition.

A.2. Integration of Message Verification Approach

In Section 4.2 the presented method for recognizing and predicting traffic situations was briefly described to be integrated into an overall system for *Car-to-X* (C2X) security. In the following this integration of the mobility data verification is detailed with respect to the overall system architecture.

The architecture of most applications for *Intelligent Transportation Systems* (ITS) was defined as a reference by the *European Telecommunications Standard Institute* (ETSI) (Intelligent Transport Systems, 2009), basically motivated by the ISO/OSI reference model (Zimmermann, 1980). The main ITS layers are *ITS Applications*, *ITS Facilities*, *ITS Network and Transport* and *ITS Access Technologies*. Furthermore, the ETSI has identified security aspects to be critical for all ITS layers in C2X communications. While cryptographic approaches are commonly used especially on network layer, the method presented in this thesis based on mobility data verification promises an easy and efficient security protection on

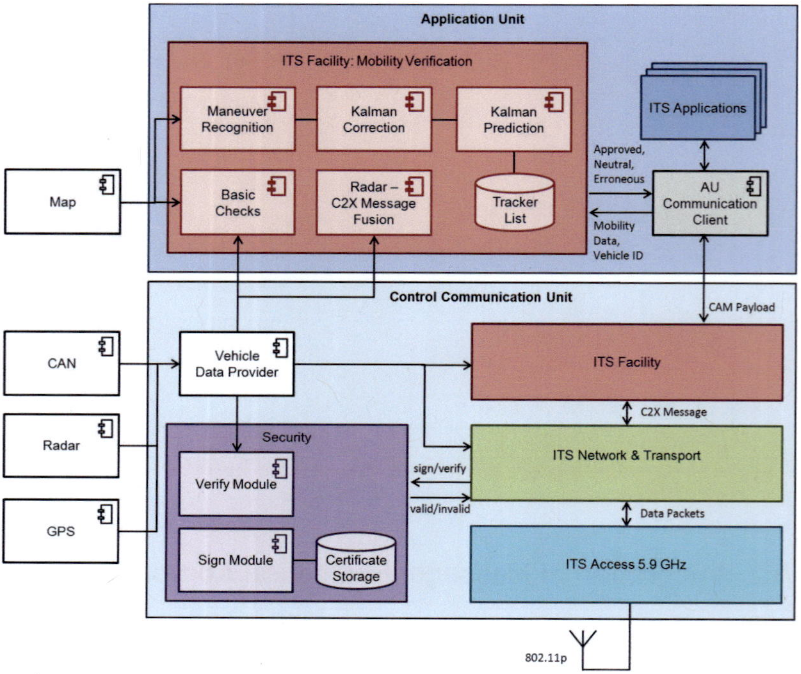

Figure A.2.: Integration of mobility data verification framework into the C2X vehicle architecture, see (Firl et al., 2013).

facility layer at a comparatively low computational cost. For an overview of different security and privacy protection approaches for the complete ITS reference model for C2X communication systems please refer to (Stübing, 2013).

The integration of the proposed mobility data verification approach into the C2X vehicle architecture is depicted in Figure A.2. This architecture was implemented using sim^{TD} vehicles (sim^{TD}, 2011) while focusing in the following on the relevant parts for message verification. The first verification stage based on a Kalman Filter was part of the official project and its field trials. However the second stage of this thesis using the maneuver prediction is also fully integrated into the framework but is not an official part of the projects implementation.

The architecture consists of two different modules: the *Control Communication Unit* (CCU) and the *Application Unit* (AU).[1] The CCU is responsible for all communication aspects between the different network layers. It is based on a 400 MHz PowerPC based on a Linux system. The AU on the other side hosts all C2X applications which are executed especially in the fields of road safety and traffic efficiency. It is running on a 2.7 GHz Dual Core system with an operating Windows Embedded. The CCU and the AU communicates with each other via Ethernet.

Incoming C2X messages are first passed through the lower access layer and are verified in the network & transport layer in a cryptographic manner. This binary verification results either in a *valid* or *invalid* message. The second verification step consists of the proposed mobility verification approach, which is triggered on facility layer. All verification stages, i.e. the basic checks, the Kalman Filter and the maneuver recognition stage (see Figure 4.9), are implemented as a separate package within the Java/OSGi bundle.

A.3. Kalman Filter for Mobility Data Verification

In the following the main equations and assumptions are presented for the usage of the Kalman Filter (KF) for mobility data verification as shortly presented in Section 4.2.2, see also (Stübing et al., 2011). As only information about position speed and heading of the vehicle is transmitted in the current

[1] Both modules form the so-called *ITS Vehicle Station* (IVS) in each equipped vehicle, compared to the fixed *ITS Central Stations* (ICS), which are each responsible for a given type of road network. In (Stübing et al., 2010) the complete sim^{TD} architecture is presented.

implementation of CAM messages, the state vector at time k of the KF is defined as:

$$x_k := (d_x, d_y, v_x, v_y) \tag{A.1}$$

Thus, the state transition model F_k is assumed to be a motion of constant velocity, i.e.:

$$F_k = \begin{pmatrix} 1 & 0 & \Delta t_k & 0 \\ 0 & 1 & 0 & \Delta t_k \\ 0 & 0 & 1 & 0 \\ 0 & 0 & 0 & 1 \end{pmatrix} \tag{A.2}$$

So the prediction step is done according to

$$\hat{x}_k = F_k \cdot \hat{x}_{k-1}^{+}. \tag{A.3}$$

The corresponding prediction error P_k, according to

$$P_k = F_k P_{k-1} F_k^T + Q_k, \tag{A.4}$$

has to be calculated and is called *system identification.* This is done offline with numerous reference traces. As a clear dependency on the current type of road, a recommend approach could be a dynamic switching between different saved system noise matrices. However, this was not implemented so far in this work and should be a topic for future work.

For the followed update stage the transition matrix H_k is used reflecting the correspondence of system state and sampled date. In this work H_k is set to the identity matrix. For weighting the new sample date with the predicted system state the Kalman Gain K_k is used:

$$K_k = P_k H_k^T (H_k P_k H_k^T + R_k)^{-1}. \tag{A.5}$$

Besides the system noise (P_k), the Kalman Gain takes the measurement noise into account, denoted as the measurement variance R_k. In contrast to P_k, the measurement noise has not to be calculated at every point in time, as incoming CAMs (sampled data) include additional information about the current accuracy of all transmitted values.

Bibliography

Aktiv (2010). Adaptive and Cooperative Technologies for the Intelligent Traffic. `http://www.aktiv-online.org/english/projects.html`.

Andrieu, C., De Freitas, N., Doucet, A., and Jordan, M. I. (2003). An Introduction to MCMC for Machine Learning. *Machine learning*, 50(1):5–43.

Archibald, R. C. (1918). Euler's Integrals and Euler's Spiral—Sometimes Called Fresnel Integrals and the Clothoide or Cornu's Spiral. *American Math Monthly*, 25:276–282.

Baader, F., Calvanese, D., McGuinness, D., Nardi, D., and Patel-Schneider, P. (2003). *The Description Logic Handbook: Theory, Implementation and Applications.* Cambridge University Press.

Bachmann, A. and Lulcheva, I. (2009). Combining Low-level Segmentation with Relational Classification. In *IEEE International Conference on Computer Vision Workshops (ICCV Workshops)*, pages 1216 –1221.

Bar-Shalom, Y., Li, X., and Kirubarajan, T. (2001). *Estimation with Applications to Tracking and Navigation: Theory Algorithms and Software.* Wiley-Interscience.

Bather, J. (2000). *Decision Theory: An Introduction to Dynamic Programming and Sequential Decisions.* Wiley Hoboken, NJ.

Baum, L. E. and Petrie, T. (1966). Statistical Inference for Probabilistic Functions of Finite State Markov Chains. *The Annals of Mathematical Statistics*, 37(6):1554–1563.

Ben-Gal, I. (2007). Bayesian Networks. *Encyclopedia of Statistics in Quality and Reliability.*

Bengio, Y. (1999). Markovian Models for Sequential Data. *Neural Computing Surveys*, 2:129–162.

Bennett, B., Cohn, A., Wolter, F., and Zakharyaschev, M. (2002). Multi-dimensional Modal Logic as a Framework for Spatio-temporal Reasoning. *Applied Intelligence*, 17(3):239–251.

Berger, J. (1985). *Statistical Decision Theory and Bayesian Analysis.* Springer-Verlag.

Berndt, D. and Clifford, J. (1994). Using Dynamic Time Warping to find Patterns in Time Series. In *AAAI-94 Workshop on Knowledge Discovery in Databases*, volume 2.

Bilmes, J. (2004). What HMMs can't do. In *Invited Paper and Lecture, ATR Workshop.*

Bilmes, J. (2006). What HMMs can do. *IEICE Transactions on Information and Systems*, 89(3):869–891.

Bishop, C. (2006). *Pattern Recognition and Machine Learning*, volume 4. Springer-Verlag.

Bißmeyer, N., Stübing, H., Schoch, E., Goetz, S., Stotz, J. P., and Lonc, B. (2011). A generic Public Key Infrastructure for securing Car-to-X Communication. In *Proceedings of the 18th World Congress on Intelligent Transportation Systems.*

Chellas, B. (1980). *Modal Logic: An Introduction.* Cambridge University Press.

Darms, M., Komar, M., and Lueke, S. (2010). Map based road boundary estimation. In *Proceedings of IEEE Intelligent Vehicles Symposium (IV)*.

Dean, T. and Kanazawa, K. (1989). A model for reasoning about persistence and causation. *Computational intelligence*, 5(2):142–150.

Drive C2X (2011). Accelerate cooperative mobility. `http://www.drive-c2x.eu/project`.

Fernández, C., Baiget, P., Roca, X., and Gonzàlez, J. (2008). Interpretation of complex situations in a semantic-based surveillance framework. *Signal Processing: Image Communication*, 23(7):554–569.

Firl, J., Stübing, H., Huss, S., and Stiller, C. (2012). Predictive maneuver evaluation for enhancement of car-to-x mobility data. In *Proceedings of IEEE Intelligent Vehicle Symposium (IV)*.

Firl, J., Stubing, H., Huss, S., and Stiller, C. (2013). MARV-X: Applying Maneuver Assessment for Reliable Verification of Car-to-X Mobility Data. *IEEE Transactions on Intelligent Transportation Systems*, 14(3):1301–1312.

Firl, J. and Tran, Q. (2011). Probabilistic maneuver prediction in traffic scenarios. In *Proceedings of European Conference on Mobile Robots (ECMR)*.

Friedman, N. and Koller, D. (2003). Being Bayesian about Network Structure. A Bayesian Approach to Structure Discovery in Bayesian Networks. *Machine learning*, 50(1):95–125.

Gerber, R. and Nagel, H.-H. (2008). Representation of Occurrences for Road Vehicle Traffic. *Artificial Intelligence*, 172:351 – 391.

Ghahramani, Z. and Jordan, M. I. (1997). Factorial Hidden Markov Models. *Machine Learning*, 29(2-3):245–273.

Gindele, T., Brechtel, S., and Dillmann, R. (2010). A Probabilistic Model for Estimating Driver Behaviors and Vehicle Trajectories in Traffic Environments. In *Proceedings of IEEE Conference on Intelligent Transportation Systems (ITSC)*.

Hahn, M., Krüger, L., and Wöhler, C. (2008). 3D Action Recognition and Long-term Prediction of Human Motion. In *Proceedings of the 6th International Conference on Computer Vision Systems (ICVS)*.

Harmelan, F., Lifschitz, V., and Porter, B. (2008). *Handbook of Knowledge Representation*, volume 1. Elsevier Science Limited.

Hasberg, C. and Hensel, S. (2008). Online-estimation of Road Map Elements using Spline Curves. In *Proceedings of International Conference on Information Fusion*.

Hasberg, C., Hensel, S., and Stiller, C. (2012). Simultaneous localization and mapping for path-constrained motion. *IEEE Transactions on Intelligent Transportation Systems*, 13(2):541–552.

Hensel, I., Bachmann, A., Hummel, B., and Tran, Q. (2010). Understanding Object Relations in Traffic Scenes. In *Proceedings of International Conference on Computer Vision Theory and Applications (VISAPP)*.

Hermes, C., Wohler, C., Schenk, K., and Kummert, F. (2009). Long-term Vehicle Motion Prediction. In *Proceedings of IEEE Intelligent Vehicles Symposium (IV)*, pages 652 –657.

Hintikka, J. (1962). *Knowledge and Belief*, volume 13. Cornell University Press Ithaca.

Hummel, B. (2009). *Description Logic for Scene Understanding at the Example of Urban Road Intersections.* PhD thesis, Universität Karlsruhe, Fakultät für Maschinenbau.

Hummel, B., Thiemann, W., and Lulcheva, I. (2007). Description Logic for Vision-based Intersection Understanding. *Proceedings of Cognitive Systems with Interactive Sensors (COGIS).*

Intelligent Transport Systems (2009). Communications Architecture. European Telecommunications Standards Institute ETSI Technical Specification TS 102 665.

Intelligent Transport Systems (2010a). European profile standard for the physical and medium access control layer of intelligent transport systems operating in the 5 GHz frequency band. ITS WG4 Std. ES 202 663, Rev. Ver. 1.1.0.

Intelligent Transport Systems (2010b). Vehicular Communications; Basic Set of Applications; Part 2: Specification of Cooperative Awareness Basic Service. European Telecommunications Standards Institute ETSI Technical Specification TS 102 637-2.

Intelligent Transport Systems (2010c). Vehicular Communications; Basic Set of Applications; Part 3: Specifications of Decentralized Environmental Notification Basic Service. European Telecommunications Standards Institute ETSI Technical Specification TS 102 637-3.

Intelligent Transportation Systems Committee (2006). IEEE Trial-Use Standard for Wireless Access in Vehicular Environments - Security Services for Applications and Management Messages. Std. 1609.2-2006.

IPG Automotive GmbH (2011). *CarMaker Reference Manual Version 3.5.*

Kammel, S., Ziegler, J., Pitzer, B., Werling, M., Gindele, T., Jagzent, D., Schröder, J., Thuy, M., Goebl, M., Hundelshausen, F., et al. (2008). Team AnnieWAY's Autonomous System for the 2007 DARPA Urban Challenge. *Journal of Field Robotics*, 25(9):615–639.

Kasper, D., Weidl, G., Dang, T., Breuel, G., Tamke, A., Wedel, A., and Rosenstiel, W. (2012). Object-Oriented Bayesian Networks for Detection of Lane Change Maneuvers. *IEEE Intelligent Transportation Systems Magazine*, 4(3):19 – 31.

Keßler, T. (2011). Situationserkennung im Straßenverkehr mittels Künstlicher Neuronaler Netze. Master's thesis, Technische Universität Darmstadt.

Koller, D., Friedman, N., Getoor, L., and Taskar, B. (2007). Graphical Models in a Nutshell. In *An Introduction to Statistical Relational Learning*. MIT Press.

Koller, D. and Pfeffer, A. (1997). Object-Oriented Bayesian Networks. In *Proceedings of Conference on Uncertainty in Artificial Intelligence (UAI)*.

Kolodner, J. (1992). An Introduction to Case-based Reasoning. *Artificial Intelligence Review*, 6:3–34.

Levinson, S. E., Rabiner, L. R., and Sondhi, M. M. (1983). An Introduction to the Application of the Theory of Probabilistic Functions of a Markov Process to Automatic Speech Recognition. *The Bell System Technical Journal*, 62:1035 –1074,.

Liebner, M., Baumann, M., Klanner, F., and Stiller, C. (2012). Driver intent inference at urban intersections using the intelligent driver model. In *Proceedings of IEEE Intelligent Vehicles Symposium (IV)*.

Liebner, M., Klanner, F., Baumann, M., Ruhhammer, C., and Stiller, C. (2013). Velocity-based driver intent inference at urban intersections in the presence of preceding vehicles. *IEEE Intelligent Transportation Systems Magazine*, 5(2):10–21.

Manna, Z. and Pnueli, A. (1992). *The Temporal Logic of Reactive and Concurrent Systems: Specification*. Springer-Verlag.

McCarthy, J. (1963a). *Programs with Common Sense*. Defense Technical Information Center.

McCarthy, J. (1963b). Situations, Actions, and Causal Laws. Technical report, Stanford Univ. Calif Dept. of Computer Science.

Meyer-Delius, D., Plagemann, C., and Burgard, W. (2009). Probabilistic Situation Recognition for Vehicular Traffic Scenarios. In *Proceedings of IEEE International Conference on Robotics and Automation (ICRA)*.

Meyer-Delius, D., Plagemann, C., von Wichert, G., Feiten, W., Lawitzky, G., and Burgard, W. (2007). A Probabilistic Relational Model for Characterizing Situations in Dynamic Multi-Agent Systems. In *Data Analysis, Machine Learning and Applications*. Springer-Verlag.

Munch, D., Jsselmuiden, J., Arens, M., and Stiefelhagen, R. (2011). High-level situation recognition using fuzzy metric temporal logic, case studies in surveillance and smart environments. In *IEEE International Conference on Computer Vision Workshops (ICCV Workshops)*, pages 882 –889.

Murphy, K. P. (2002). *Dynamic Bayesian Networks: Representation, Inference and Learning*. PhD thesis, University of California.

Øhrstrøm, P. and Hasle, P. (1995). *Temporal Logic: From Ancient Ideas to Artificial Intelligence*, volume 57. Springer-Verlag.

Omerbegovic, S. (2012). Situation-adaptive Vehicle Trajectory Prediction. Master's thesis, Technische Universität Darmstadt.

Omerbegovic, S. and Firl, J. (2013). Situation-adaptive Vehicle Trajectory Prediction. In *Proceedings of Automotive meets Electronics (AmE)*.

Perner, P. (2007). *Case-Based Reasoning on Images and Signals*, volume 73. Springer-Verlag.

Rabiner, L. (1989). A Tutorial on Hidden Markov Models and Selected Applications in Speech Recognition. *Proceedings of the IEEE*, 77(2):257–286.

Richardson, M. and Domingos, P. (2006). Markov Logic Networks. *Machine Learning*, 62(1-2):107–136.

Russell, S. and Norvig, P. (2009). *Artificial Intelligence: A Modern Approach*. Prentice hall.

Schäfer, K. H. (1996). *Unscharfe zeitlogische Modellierung von Situationen und Handlungen in Bildfolgenauswertung und Robotik*. PhD thesis, Fakultät für Informatik, Universität Karlsruhe (TH).

Schubert, R. and Wanielik, G. (2011). A Unified Bayesian Approach for Object and Situation Assessment. *IEEE Intelligent Transportation Systems Magazine*, 3(2):6 –19.

Selinger, M. (2009). *Klassifikation von Verkehrssituationen für ein aktives Sicherheitssystem mit einem Neuronalen Netz*. PhD thesis, Universität Karlsruhe (TH).

sim^{TD} (2011). Sichere Intelligente Mobilität Testfeld Deutschland. `www.simtd.de`.

Smyth, P. (1997). Belief Networks, Hidden Markov Models, and Markov Random Fields: A Unifying View. *Pattern Recognition Letters*, 18(11):1261–1268.

Stiller, C., Kammel, S., Lulcheva, I., and Ziegler, J. (2008). Probabilistische Methoden in der Umfeldwahrnehmung Kognitiver Automobile. *at-Automatisierungstechnik*, 56(11):563–574.

Stübing, H. (2013). *Multilayered Security and Privacy Protection in Car-to-X Networks: Solutions from Application down to Physical Layer*. Springer-Verlag.

Stübing, H., Bechler, M., Heussner, D., May, T., Radusch, I., Rechner, H., and Vogel, P. (2010). sim^{TD}: A Car-to-X System Architecture for Field Operational Tests. *IEEE Communications Magazine*, 48(5):148–154.

Stübing, H., Firl, J., and Huss, S. (2011). A Two-Stage Verification Process for Car-to-X Mobility Data based on Path Prediction and Probabilisitc Maneuver Recognition. In *Proceedings of IEEE Vehicle Networking Conference (VNC)*.

Stübing, H., Jaeger, A., Bißmeyer, N., Schmidt, C., and Huss, S. A. (2010). Verifying Mobility Data under Privacy Considerations in Car-to-X Communication. In *Proceedings of 17th World Congress on Intelligent Transportation Systems(ITS)*.

Vacek, S., Gindele, T., Zollner, J., and Dillmann, R. (2007). Situation Classification for Cognitive Automobiles using Case-based Reasoning. In *Proceedings of IEEE Intelligent Vehicles Symposium (IV)*.

Vlachos, M., Gunopulos, D., and Das, G. (2004). Rotation Invariant Distance Measures for Trajectories. In *Proceedings of ACM SIGKDD International Conference on Knowledge Discovery and Data Mining*. ACM.

Vlachos, M., Kollios, G., and Gunopulos, D. (2005). Elastic Translation Invariant Matching of Trajectories. *Machine Learning*, 58(2):301–334.

Winner, H., Hakuli, S., and Wolf, G. (2009). *Handbuch Fahrerassistenzsysteme: Grundlagen, Komponenten und Systeme für aktive Sicherheit und Komfort*. Vieweg + Teubner.

Zhang, Z., Huang, K., and Tan, T. (2006). Comparison of Similarity Measures for Trajectory Clustering in Outdoor Surveillance Scenes. In *Proceedings of International Conference on Pattern Recognition (ICPR)*.

Zimmermann, H. (1980). OSI Reference Model–The ISO Model of Architecture for Open Systems Interconnection. *IEEE Transactions on Communications*, 28(4):425–432.

Zucchini, W. and MacDonald, I. L. (2009). *Hidden Markov Models for Time Series: An Introduction using R*. Monographs on statistics and applied probability;. CRC Press.

Schriftenreihe
Institut für Mess- und Regelungstechnik
Karlsruher Institut für Technologie
(1613-4214)

Die Bände sind unter www.ksp.kit.edu als PDF frei verfügbar
oder als Druckausgabe bestellbar.

Band 001 Hans, Annegret
Entwicklung eines Inline-Viskosimeters auf Basis eines magnetisch-induktiven Durchflussmessers. 2004
ISBN 3-937300-02-3

Band 002 Heizmann, Michael
Auswertung von forensischen Riefenspuren mittels automatischer Sichtprüfung. 2004
ISBN 3-937300-05-8

Band 003 Herbst, Jürgen
Zerstörungsfreie Prüfung von Abwasserkanälen mit Klopfschall. 2004
ISBN 3-937300-23-6

Band 004 Kammel, Sören
Deflektometrische Untersuchung spiegelnd reflektierender Freiformflächen. 2005
ISBN 3-937300-28-7

Band 005 Geistler, Alexander
Bordautonome Ortung von Schienenfahrzeugen mit Wirbelstrom-Sensoren. 2007
ISBN 978-3-86644-123-1

Band 006 Horn, Jan
Zweidimensionale Geschwindigkeitsmessung texturierter Oberflächen mit flächenhaften bildgebenden Sensoren. 2007
ISBN 978-3-86644-076-0

Band 007 Hoffmann, Christian
Fahrzeugdetektion durch Fusion monoskopischer Videomerkmale. 2007
ISBN 978-3-86644-139-2

Band 008 Dang, Thao
Kontinuierliche Selbstkalibrierung von Stereokameras. 2007
ISBN 978-3-86644-164-4

Band 009 Kapp, Andreas
Ein Beitrag zur Verbesserung und Erweiterung der Lidar-Signalverarbeitung für Fahrzeuge. 2007
ISBN 978-3-86644-174-3

Band 010 Horbach, Jan
Verfahren zur optischen 3D-Vermessung spiegelnder Oberflächen. 2008
ISBN 978-3-86644-202-3

Band 011 Böhringer, Frank
Gleisselektive Ortung von Schienenfahrzeugen mit bordautonomer Sensorik. 2008
ISBN 978-3-86644-196-5

Band 012 Xin, Binjian
Auswertung und Charakterisierung dreidimensionaler Messdaten technischer Oberflächen mit Riefentexturen. 2009
ISBN 978-3-86644-326-6

Band 013 Cech, Markus
Fahrspurschätzung aus monokularen Bildfolgen für innerstädtische Fahrerassistenzanwendungen. 2009
ISBN 978-3-86644-351-8

Band 014 Speck, Christoph
Automatisierte Auswertung forensischer Spuren auf Patronenhülsen. 2009
ISBN 978-3-86644-365-5

Band 015 Bachmann, Alexander
Dichte Objektsegmentierung in Stereobildfolgen. 2010
ISBN 978-3-86644-541-3

Band 016 Duchow, Christian
Videobasierte Wahrnehmung markierter Kreuzungen mit lokalem Markierungstest und Bayes'scher Modellierung. 2011
ISBN 978-3-86644-630-4

Band 017 Pink, Oliver
Bildbasierte Selbstlokalisierung von Straßenfahrzeugen. 2011
ISBN 978-3-86644-708-0

Band 018 Hensel, Stefan
Wirbelstromsensorbasierte Lokalisierung von Schienenfahrzeugen in topologischen Karten. 2011
ISBN 978-3-86644-749-3

Band 019 Carsten Hasberg
Simultane Lokalisierung und Kartierung spurgeführter Systeme. 2012
ISBN 978-3-86644-831-5

Band 020 Pitzer, Benjamin
Automatic Reconstruction of Textured 3D Models. 2012
ISBN 978-3-86644-805-6

Band 021 Roser, Martin
Modellbasierte und positionsgenaue Erkennung von Regentropfen in Bildfolgen zur Verbesserung von videobasierten Fahrerassistenzfunktionen. 2012
ISBN 978-3-86644-926-8

Band 022 Loose, Heidi
Dreidimensionale Straßenmodelle für Fahrerassistenzsysteme auf Landstraßen. 2013
ISBN 978-3-86644-942-8

Band 023 Rapp, Holger
Reconstruction of Specular Reflective Surfaces using Auto-Calibrating Deflectometry. 2013
ISBN 978-3-86644-966-4

Band 024 Moosmann, Frank
Interlacing Self-Localization, Moving Object Tracking and Mapping for 3D Range Sensors. 2013
ISBN 978-3-86644-977-0

Band 025 Geiger, Andreas
Probabilistic Models for 3D Urban Scene Understanding from Movable Platforms. 2013
ISBN 978-3-7315-0081-0

Band 026 Hörter, Marko
Entwicklung und vergleichende Bewertung einer bildbasierten Markierungslichtsteuerung für Kraftfahrzeuge. 2013
ISBN 978-3-7315-0091-9

Band 027 Kitt, Bernd
Effiziente Schätzung dichter Bewegungsvektorfelder unter Berücksichtigung der Epipolargeometrie zwischen unterschiedlichen Ansichten einer Szene. 2013
ISBN 978-3-7315-0105-3

Band 028 Lategahn, Henning
Mapping and Localization in Urban Environments Using Cameras. 2013
ISBN 978-3-7315-0135-0

Band 029 Tischler, Karin
Informationsfusion für die kooperative Umfeldwahrnehmung vernetzter Fahrzeuge. 2014
ISBN 978-3-7315-0166-4

Band 030 Schmidt, Christian
Fahrstrategien zur Unfallvermeidung im Straßenverkehr für Einzel- und Mehrobjektszenarien. 2014
ISBN 978-3-7315-0198-5

Band 031 Firl, Jonas
Probabilistic Maneuver Recognition in Traffic Scenarios. 2014
ISBN 978-3-7315-0287-6